Advances in Urethane Science and Technology

Advances in Urethane Science and Technology

VOLUME 13

EDITED BY:

Kurt C. Frisch

Daniel Klempner

POLYMER TECHNOLOGIES, INC.
UNIVERSITY OF DETROIT

CRC Press
Taylor & Francis Group
Boca Raton London New York

CRC Press is an imprint of the
Taylor & Francis Group, an **informa** business

Advances in Urethane Science and Technology–Volume 13

First published 1996 by Technomic Publishing Company, Inc.

Published 2020 by CRC Press
Taylor & Francis Group
6000 Broken Sound Parkway NW, Suite 300
Boca Raton, FL 3487-2742

ISBN-13: 978-1-56676-410-0 (hbk)

Visit the Taylor & Francis Web site at
http://www.taylorandfrancis.com

and the CRC Press Web site at
http://www.crcpress.com

ISSN No. 0044-6378

Table of Contents

PU Automotive Carpet Composites for Vehicle Interior Noise Control: Physical Basis of Behaviour

A. Cunningham
ICI Polyurethanes Research Centre
Everslaan 45
B-3078 Everberg, Belgium

N. C. Hilyard
Materials Research Institute
Division of Applied Physics
Sheffield Hallam University
Pond Street
Sheffield, S1 1WB, UK

ABSTRACT

Moulded all MDI PU foam backed floor pan/bulkhead carpet composites for vehicle interior noise control offer advantages over alternative systems and are now well established in the automotive sector. As part of our development programme a systematic theoretical and experimental study of their acoustic insulation/isolation behaviour has been carried out. Noise enters a vehicle interior through two routes, air and structure, so both situations have been addressed. Early theoretical models for light weight multi-element systems concentrated on aircraft panels with rigid fibre cores. These models are adapted to suit the needs of the current situation in which the core is viscoelastic porous rather than rigid porous. To focus attention on the acoustic behaviour of the composite, the parameter measured and modelled is the composite Insertion Loss, IL. Modelling and measurement show that for structure and airborne noise at low acoustic frequencies (~ 100 Hz) the IL of the composite is positive, indicating an attenuation of the transmitted noise, but it decreases as the frequency increases to a resonance trough. At the trough frequency the IL value may be negative, i.e. the noise transmission with composite in place is larger than that with bare metal. System resonance plays a major role in governing noise insulation behaviour and this is controlled by design and material parameters. Modelling identifies quantitatively the main factors affecting the IL response; e.g. the surface mass of the steel panel and heavy layer and the thickness, modulus, loss factor and airflow resistivity of the core foam. For airborne noise the quantitative agreement between modelling and measurement is good, except at system resonance where the predicted IL is about 3–4 dB lower than that measured using a test rig with simple geometry. The reasons for this are discussed. The model for vibration excited noise is approximate and although the predictions agree quantitatively with some aspects of measured behaviour it requires further development.

1. INTRODUCTION

Over the past decade the reduction of vehicle interior noise through the incorporation of moulded polyurethane foam backed carpet composites has been the focus of development by suppliers, moulders and vehicle builders [1–5]. Many forces have driven these developments. They include cost, weight reduction, improvements in online efficiency and enhanced acoustic performance. With respect to the last factor it has been shown [5,6] that typically a 3–5 dBA reduction of interior noise level relative to bare metal can be gained by fitting moulded polyurethane floor pan/bulkhead composites. Because of the keen interest in the acoustic performance of these systems a broadly based theoretical and experimental study of their behaviour has been carried out by the authours and reported in the literature [7–11].

This report reviews the results of these investigations and gives a more detailed description of our modelling approach. Background to the investigation is provided by considering the sources and origin of vehicle interior noise. The report then considers theoretical models used to explain the acoustic behaviour of multi-element structures and methods used for their evaluation. This is followed by the presentation of experimental data for vibration insertion loss and airborne insertion loss behaviour measured using a relatively small scale laboratory based test rig. Measured behaviour is reconciled with model predictions.

2. BACKGROUND

Vehicle interior noise originates from several exterior sources, the main contributions coming from engine and drive train, road/tyre, and wind. Each of these sources has specific characteristics in terms of level and frequency content. The noise enters the interior through two pathways, air and vehicle structure. The relative balance between airborne noise and structure (vibration) borne noise is illustrated in Figures 1 and 2.

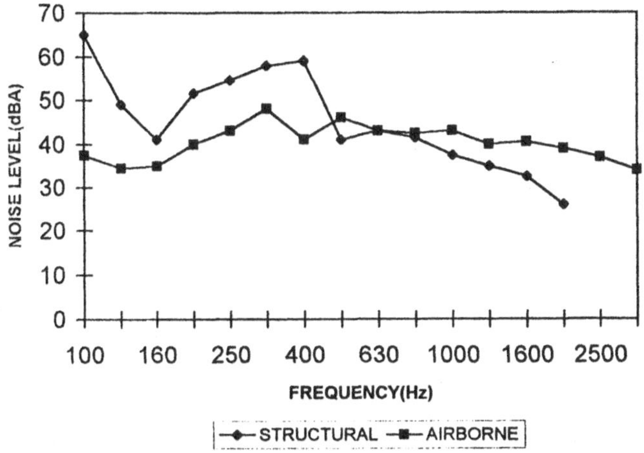

Figure 1. Contributions of airborne and structure borne noise to overall engine sourced interior noise (adapted from Reference [12]).

Figure 1 shows the contributions to the overall interior noise level for engine sourced noise in a laboratory tested passenger vehicle [12]. (Here, and in subsequent chart presentations of experimentally measured noise levels, the noise is A-weighted and plotted as 1/3 octave band levels.) It is seen from Figure 1 that below the 1/3 octave band centred at 500 Hz the contribution from structure borne noise originating from the engine is larger than the airborne noise. Above 500 Hz it is the reverse, airborne noise being greater than structure borne noise (an

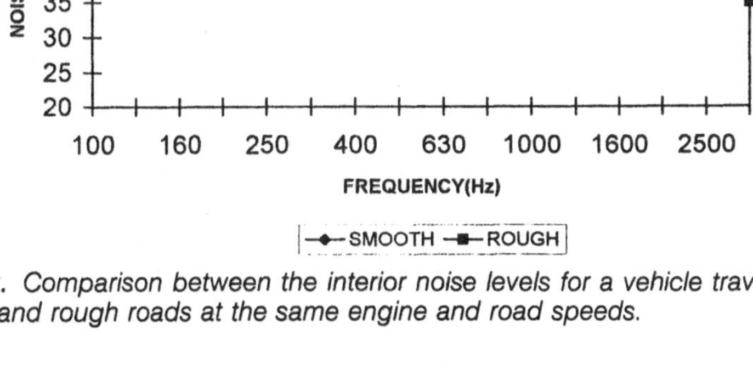

Figure 2. Comparison between the interior noise levels for a vehicle travelling on smooth and rough roads at the same engine and road speeds.

observation also made in Reference [13]). The data presented in Figure 2 shows interior noise levels for a vehicle travelling on-road for two road surfaces, smooth and rough, under the same conditions of speed and gear, and measured at the same position within the interior compartment [6]. Since other conditions remain the same we can interpret the difference as being due primarily to the road/tyre exterior noise source, the rough road surface producing higher levels of mechanical vibration than the smooth surface. It is seen that below the 1/3 octave band centred at 630 Hz, the level of noise for the rough surface is greater than that for the smooth surface but above it is less, which is in agreement with Reference [12].

It is concluded that the contribution to vehicle interior noise level from the airborne and structure borne sources depends on the frequency and that below ~500–630 Hz structure borne noise predominates whereas above this frequency airborne noise predominates. The on-road data indicates that at frequencies in the range 100–300 Hz the difference between band levels is ~5 dBA.

Although noise is transmitted through two pathways it radiates into the interior from numerous locations; floor pan/bulkhead, roof, A/B pillars, front and rear screens, etc. It is then conditioned by cavity resonances (boom noise) and noise absorption (see for example References [13,14]). The relative importance of these interior source locations are difficult to quantify but for the front end of a passenger vehicle Gahlau [4] divided them into two (1) floor pan/bulkhead and (2) "others." Data derived from his article shows that below the 1/3 octave band centred at 160 Hz the contributions to the overall noise from the panel/bulkhead system and "other" are about the same, i.e. ~3 dB down on the total. This means that if the noise from the panel/bulkhead were eliminated completely the interior noise levels in these bands would be reduced by ~3 dB. Above 160 Hz the relative contribution from the floor pan/bulk head system increases whereas that from "other" sources decreases such that at ~1 kHz the contribution from the floor-pan/bulkhead is only 1 dB down from the total whereas that from the "other" sources is ~8 dB down. If the noise from the panel/bulkhead system were eliminated under these circumstances the interior band noise level would be reduced by ~8 dB. It is shown below that typically the insertion loss of a polyurethane foam backed carpet system for both airborne and structure borne noise is >12–15 dB at these band frequencies so it is concluded that in a "typical" vehicle fitted with a "typical carpet system" it is the "other" sources which make the major contribution to interior noise at frequencies > about 1 kHz. Thus from a practical viewpoint our interest is with both airborne and structure borne noise and mainly, but not exclusively, with the acoustic insulation/isolation behaviour of the carpet composite at frequencies below 1 kHz.

3. CHARACTERISATION OF COMPOSITE PERFORMANCE

The behaviour of a noise attenuating composite is commonly measured in terms of the Transmission Loss, TL, or in Europe the Sound Reduction Index, SRI. This is represented by the parameter R defined by the equation

$$R(\text{dB}) = 10 \log (1/\tau) = 20 \log |p_i/p_t| \quad (1)$$

where τ is the transmission coefficient of the structure, p_i the pressure amplitude in the sound field on the incident side of the structure and p_t the pressure amplitude on the receiving, or transmitting, side. Gahlau [4] has presented data for the Transmission Loss of carpet composites determined from in-car laboratory tests. Typical data for the front end of the vehicle interior are illustrated in Figure 3(a) which compares the behaviour of a low density foam core + septum with that of a viscoelastic (VE) foam core + septum where the area mass of the septum was 5 kg/m^2. The low density foam was not defined but the VE foam had modulus $\sim 1 \times 10^5$ Pa and loss factor >0.2. The thickness of the core was 20 mm and the thickness of the steel panel 1.0 mm. Below 315 Hz the response is approximately flat and above it increases strongly with increasing frequency. In the frequency range shown the panel would be expected to give a TL response increasing monotonically with increasing frequency. Thus the peaks and troughs seen in the response can be attributed to the effect of the carpet composite.

At low acoustic frequencies the added TL due to the composite is relatively small compared with that of the steel panel so the measurement strategy widely adopted is to remove the effect of the panel. This enables us to focus on the effects of composite design variables. The approach is illustrated as follows.

The normal incidence Transmission Loss R_0 for a limp mass panel (the mass law) can be written as [15]

$$R_0 = 10 \log [1 + (\omega m_2/2\varrho_0 c_0)^2] \quad (2)$$

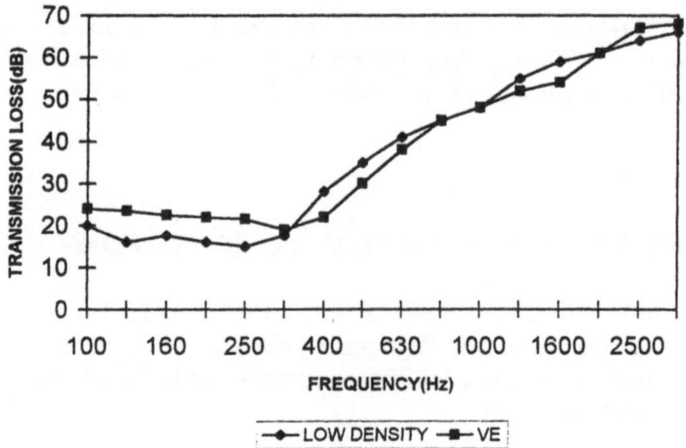

(a)

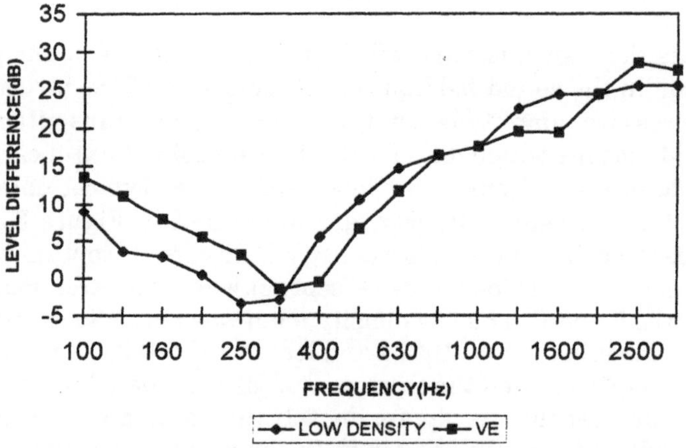

(b)

Figure 3. *(a) Comparison of the transmission loss of a floor pan/bulkhead system with two polyurethane foam backed carpet composites [4]; (b) Level difference for the two carpet systems calculated using the field incidence limp panel mass law.*

where

ϱ_0 = density of air
c_0 = speed of sound in air
m_1 = surface mass of panel

For the in-car tests of Gahlau the acoustic field would not be incident normally on the exterior of the floor pan/bulkhead system so we take the field incidence law applicable to low acoustic frequencies, which from V'Er and Holmer [15] is

$$R_{field}\,(\text{dB}) = R_0 - 5 \tag{3}$$

Subtracting the field incident dB Transmission Loss of the floor pan given by (2) and (3) from the measured TL (dB) data of Gahlau gives the semi-empirical in-car Level Difference (LD) provided by the carpet composite. The results of this procedure are illustrated in Figure 3(b) for the two foam cores. It is seen that at 100 Hz the LD values are in the range 8–14 dB and as the frequency increases the LD values decrease to a trough, become negative and then increase. Gahlau attributed the trough to "the resonance of the mass-spring system," i.e. the panel + flexible foam core + septum. The conclusions from this analysis indicate that the resonance trough frequency and the LD at the trough are dependent on the physical properties of the foam core; the VE foam core giving a higher resonance frequency and smaller trough than the low density foam.

The aim of our work has been to explain quantitatively the reasons for observed differences in behaviour and from this develop foam systems which can be tailored to meet the needs of a vehicle with a specific noise input signature. To do this our theoretical and experimental strategy has been to examine and probe the factors that influence the Level Difference provided by the composite. The procedure used is illustrated in Figure 4. Figure 4(a) shows the bare metal panel with incident noise level L(0) and level L(1) on the receiving side. For the purpose of explaining the insertion loss of the composite, Figure 4(b), the tufted carpet is subsumed into the septum; the carpet merely adding

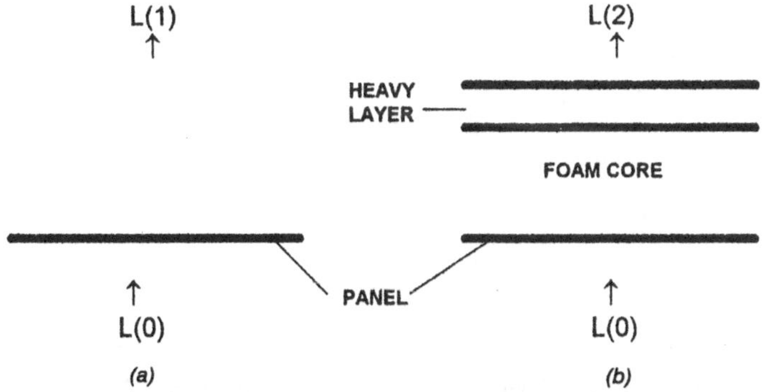

Figure 4. Insertion Loss measurement concept.

mass. (In-car, however the effect of the carpet cannot be ignored since it will influence the overall vehicle interior noise level through the process of acoustic absorption.) In the case of the panel + composite the incident noise level is assumed to be the same as before, i.e. L(0), and the level on the receiving side L(2). The Level Difference is the difference in noise levels L(1)–L(2) on the receiving side without and with the carpet composite in place all other conditions remaining the same. Under "ideal" laboratory conditions, unlike a vehicle interior, there are no "other" sources and we define the Insertion Loss of the composite, IL, as

$$\text{IL (dB)} = \text{L(1)} - \text{L(2)} \tag{4}$$

and this is the parameter modelled and measured in our work. A quantitative indication of the behaviour to be expected can be obtained from the data given by Berthevas et al. [2]. The PU composite was bonded to a 1 m × 1 m steel panel test panel mounted vertically in a transmission suite. The data are given in Table 8 of the original article and are presented in the form of a chart in Figure 5. It is seen that the IL values fall to about zero dB or below at the resonance trough (315 Hz) and then increases strongly above resonance.

Product acceptance performance is often expressed in terms of minimum IL values at predetermined 1/3 octave band centre frequencies

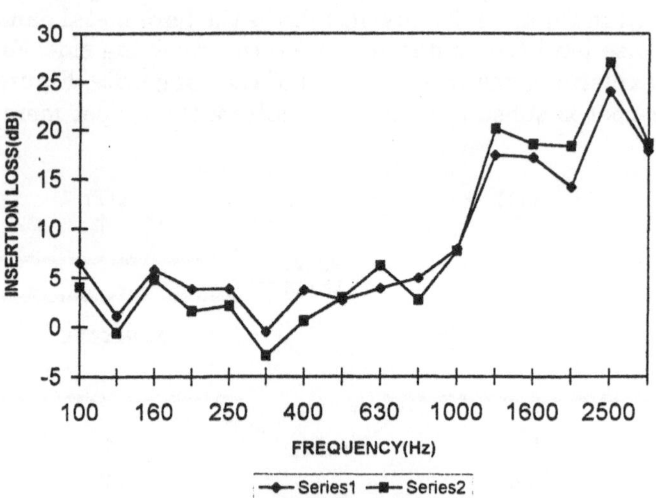

Figure 5. Insertion Loss of two PU foam cored composites bonded to steel panel [2].

Table 1. Example of lower limit IL values for automotive carpet systems.

Frequency (Hz)	IL_{limit} (dB)
500	2
1000	22
3000	38
5000	40

measured using a specified test rig. Limit values set in the past are given in Table 1.

4. THEORETICAL BASIS

The purpose of the modelling work was two fold: (1) to identify the independently measurable parameters which control the behaviour of the composite (such as the surface mass of the heavy layer (septum) and the thickness, stiffness and damping of the foam core) and (2) establish the credibility of experimental data produced by the laboratory test rig used in our studies which is described below. Structures can support four types of mechanical wave; longitudinal, shear, torsional and bending. With thin panels the most important with respect to the radiation of noise into the vehicle interior are the bending waves. Observed differences between structure borne noise and airborne noise behaviour, described below, might be attributed to the processes by which the noise is transmitted to the vehicle interior.

Panel bending waves resulting from point or line vibration excitation, propagate away from the point of excitation with velocity c_B and wavelength λ_B and radiate noise on both sides of the panel. In a loss free system the waves would propagate unattenuated but in a semi-ideal system the amplitude of the waves would decrease through energy losses in the material of the panel and radiation loss Reference [16a]. The vibration close to the point of excitation is called the near vibration field. The bending wavelength in a panel with surface mass m_1 and bending stiffness B per unit width is given by [16b]

$$\lambda_B = 2\pi[B/(\omega^2 m_1)]^{1/4}$$

and this varies inversely as the square root of the frequency. Since the wavelength in the air on either side of the panel, $\lambda_0 = c_0/f$, varies inversely with respect to the frequency there exists a unique frequency for which the wavelength in air equals the wavelength in the panel [16b]. This frequency is called the critical, or coincidence, frequency and is given by

$$f_c = (c_0^2/2\pi)[m_1/B]^{1/2}$$

where m_1 is the surface mass (kg/m²) and B (Nm) the bending stiffness per unit width of the panel. At coincidence radiation/transmission is enhanced so that in practical systems there is a dip, or trough, in the noise insulation response. For a 1 mm thick steel panel, such as that used in our test rig, and close to that used in most passenger vehicles, the coincidence frequency is ~12 kHz [16b].

Structure borne noise excites the vehicle interior shell at points or lines. However incident airborne noise excites the panel over a distributed area and the panel subsequently radiates on both sides. The major difference between vibration excitation and airborne noise excitation is point/line, with attenuated wave propagation in the panel, and distributed, which excites the panel over the entire surface. Coincidence, as described above for vibration excitation, occurs with incident airborne noise so when the airborne wavelength is the same as the panel bending wave we would expect an enhancement in transmission and a trough in IL response.

Theoretical approaches used to explain the noise behaviour of light weight multi-element composites with porous cores are divided into two categories according to complexity; continuum [17] and micromechanical [18–20]. Although the micromechanical approaches have a sound physical basis their development is complex and they are difficult to apply to "real" situations. According to Bolton [20] interaction between the radiated noise fields due to vibration and airborne excitation occurs only when the sources are coherent. In the model development given below we assume that this is not the case.

4.1 Structure Borne Noise

Structure borne noise originates from two main sources: road vibration (via suspension) and engine vibration (via mountings). As explained above noise energy penetrates the interior compartment by various routes and evidence shows that its major influence on interior noise quality is at frequencies < ~500 Hz. The primary factors concerning the ingress of vibration borne noise are (a) the interior source, e.g. floor pan/bulkhead, roof, etc., and (b) the frequency spectrum. A physical insight to factors governing structure borne noise is obtained from the vibration behaviour of finite clamped panels.

4.1.1 Vibration of Clamped Panels

In finite panels a bending wave propagating along the panel is reflected at the edges. Interference between reflected and incident waves produces standing wave patterns which may result in large amplitude transverse panel motion. For a "simply supported" panel, i.e. one in which the displacement and curvature at the boundaries are zero, the mode frequencies are predicted to be [15,16c]

$$f_{m,n} = (\pi/2)(B/m_s)^{1/2}[(m/l_x)^2 + (n/l_y)^2] \tag{5}$$

where

B = bending stiffness of the panel per unit width (Nm)
m_s = surface mass of the panel (kg/m^2)
l_x,l_y = side length of panel in x and y directions
m,n = 1,2,3 . . .

The panel deformation shape for a particular value of m,n is called a normal mode. The shapes given by finite element analysis for the first three modes $f_{1,1}$, $f_{1,2}$, and $f_{2,2}$ of a C-C-C-C (or 4C) panel are illustrated in Figure 6. Lightly damped panels of finite dimensions can support

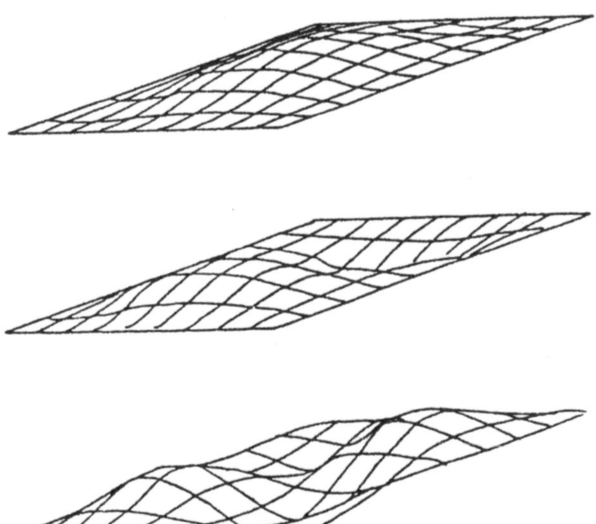

Figure 6. Panel displacement of the first three bending modes (1,1), (1,2), and (2,2) in a square 4C panel given by finite element analysis.

many modes and this leads to a high modal density, i.e., many closely spaced, sometimes overlapping, resonances. The measured response for the point excited bare metal 4C panel, 0.87 m × 0.87 m × 1 mm used in the test rig described below is shown for frequencies in the range 100 Hz–1 kHz in Figure 7(a). The effect of an HR PU foam core (d = 20 mm, ϱ_f = 80 kg/m^3) placed on top of the panel is illustrated in Figure 7(b). It is seen that it reduces both the intensity of the modal resonances and the background vibration level but does not significantly influence the mode frequencies. The effect of placing a 5.9 kg/m^2 nonbonded high density EVA heavy layer on top of the core is illustrated in Figure 7(c) which shows that the effect of the core layer is enhanced and the modal resonances almost, but not completely, suppressed.

The damping of the panel in the test rig may be attributed to three factors: (1) the clamped boundaries, (2) the absorption and dissipation of panel energy within the foam, and (3) slip (friction) at the panel/foam interface. Radiation damping is considered to be unimportant in this situation [16a]. The relative importance of the two mechanisms involving the foam core has yet to be established. It is difficult to assess the

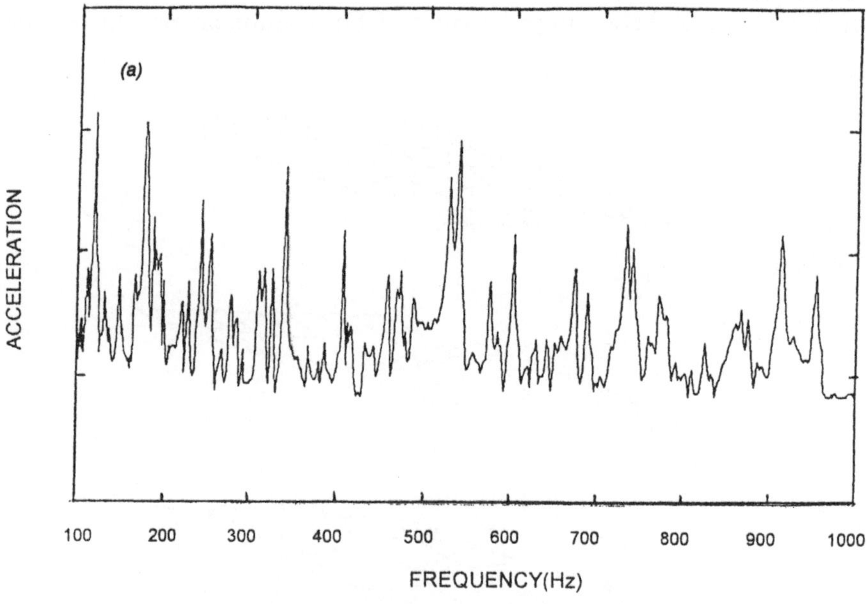

Figure 7. Swept sine frequency response for the panel of the test rig in the multimode region. (a) Bare metal.

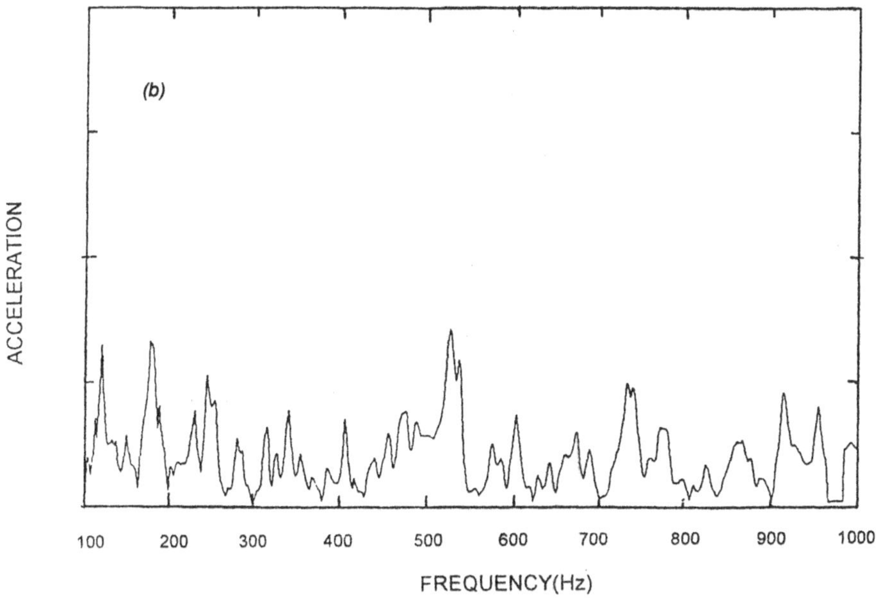

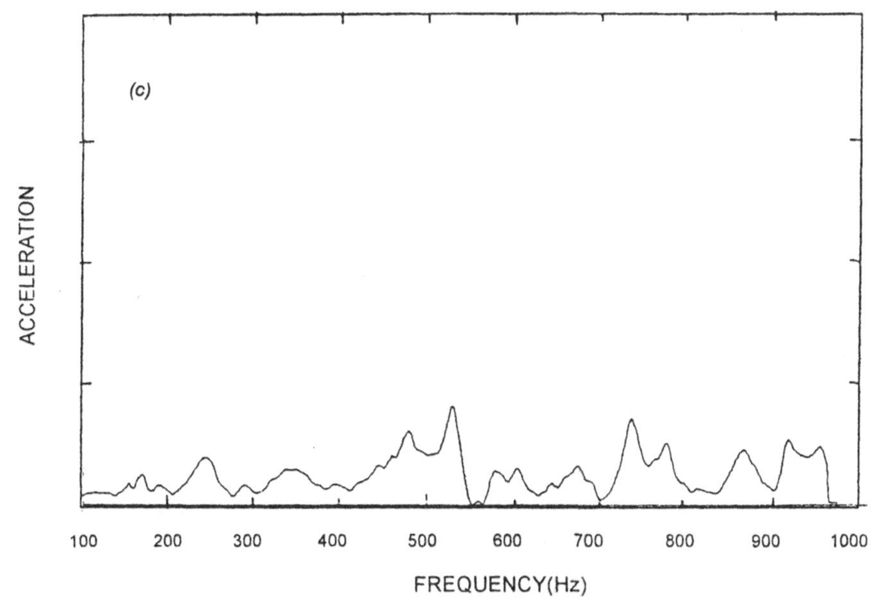

Figure 7 (continued). *Swept sine frequency response for the panel of the test rig in the multi-mode region. (b) with 20 mm thick HR foam core and (c) with foam core + 5.9 kg/m² septum.*

overall effect of damping on radiated noise because the sound power radiated from a structure depends not only on the mean square velocity but also the wavelengths of the structure and these are different for different modes of vibration. In the case of a single resonance mode damping of the structure reduces the mean square velocity in proportion to $1/\eta^2$, where η is the loss factor of the structure, and so results in a considerable reduction of the radiated noise. However in the multi-mode region the situation is more complicated.

4.1.2 Sound Radiated by Clamped Panels

We are concerned with the noise radiated by a rectangular 4C panel with point vibration excitation, as in the test rig. The mechanical input power generates bending waves and in the multi-mode region the bending wave field can be divided into two parts: (a) the field created directly by the exciting force F, which propagates away from the point of excitation, called the vibration near field, and (b) the reverberant field created by reflection at the panel boundaries. Under these conditions the acoustic power P radiated by a panel excited by a point harmonic force F is from [16e]

$$P = [\varrho_0 c_0 k^2 F^2/2\pi\omega^2 m_s^2]\{1 + U/2Sk_B\eta\} \tag{6}$$

where the first term in braces represents the power resulting from the near field and the second from the reverberant field. In this equation

$k = 2\pi/\lambda_0$ = wave number in air
m_s = surface mass of panel
U = length of panel boundary
S = area of panel
$k_B = 2\pi/\lambda_B$ = bending wave number in panel
η = effective loss factor of panel

As Cremer and Heckl state, "for plates with very low damping an increase in the loss factor causes a proportional reduction of radiated sound; if, however the loss factor is already larger than

$$\eta_{limit} = U/2Sk_B \tag{7}$$

a further increase has no effect."

The value of $k_B = 2\pi/\lambda_B$ can be calculated in several ways. The most

Table 2. Predicted optimum loss factors for a 1 m × 1 m × 1 mm 4C steel panel.

f/Hz	λ_B	η_{limit}
100	0.3	0.12
1000	0.1	0.04
5000	0.04	0.016

convenient is from the critical frequency f_c which is related to the excitation frequency f by

$$\lambda_B = c_0/(ff_c)^{1/2} \tag{8}$$

Optimum panel loss factors for a 1 m × 1 m × 1 mm panel, $f_c \sim 12$ kHz, calculated from Equations (7) and (8) are given in Table 2.

The loss factors of damped panels can be measured in several ways, e.g., mode analysis, reverberation decay and the power injection method (PIM). Plunt [21] has critically compared these procedures and concludes that mode analysis is limited to low order modes, the reverberation decay method may emphasise lightly damped modes at the expense of moderately and highly damped modes and that the preferred procedure is the power injection method. Unfortunately few experimental data obtained using PIM are available for discussion so it is necessary to rely on values of panel loss factor determined by the decay method [23].

A miniature accelerometer was attached to the 4C panel in the test rig and the panel excited using a hammer. The vibration decay was measured in 1/3 octave bands by means of the RTA set in transient mode. The loss factor was calculated from $\eta = 1.1/f_n T_{30}$ where f_n is the band centre frequency and T_{30} the time for a 30 dB decay of the signal level. Typical behaviour measured using this procedure is given in Figures 8(a) and (b). Figure 8(a) shows the loss factor of the bare metal panel and the panel with an unbonded foam core layer placed on top and taped around the edges of the frame. At low frequencies the presence of the core enhances the loss factor by a significant amount, the loss factor for the panel + core at 200 Hz being ~0.02 and that for the bare panel ~0.003. This conforms to the subjective change in the modal responses illustrated in Figure 7. However the amount of enhancement decreases with increasing frequency. The reason for this is not known, it might be associated with the measurement procedure. Figure 8(b) shows that increasing the surface mass of the septum produces only a small increase of effective loss factor. Although in this il-

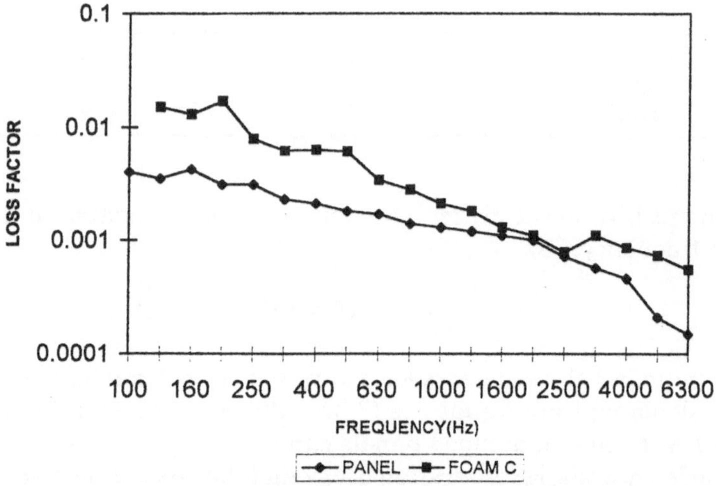

(a)

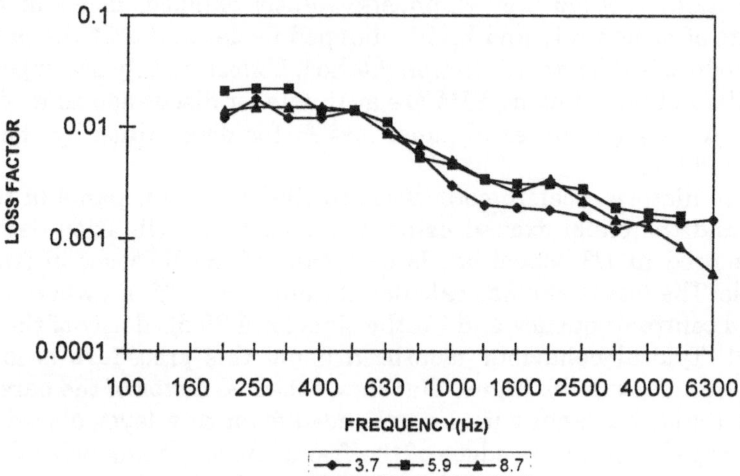

(b)

Figure 8. *Panel loss factor as a function of frequency. (a) Comparison of bare metal and bare metal + non-bonded foam core; (b) Influence of septum mass.*

lustration there is little differentiation between data for the different surface masses the overall indication from several tests indicated that higher area mass septums produced a small enhancement of loss factor. The scatter in measured data does not warrant more detailed analysis.

Two observations are made: (1) A foam core placed on the panel should affect the noise radiated by the panel, and (2) the measured value of loss factor of a typical panel + core + septum system is less than the optimum. However, as commented above, the values measured using the decay method may be less than those which obtain in reality.

Assuming the foam core alone does not add mass or stiffness to the panel + core system (and doesn't affect the radiation efficiency) Equation (6) shows that the ratio of the acoustic power P_{10} radiated by the bare panel to the power P_1 radiated with the core in place can be expressed as

$$P_{10}/P_1 = \{1 + (U/2S)/k_B\eta_{10}\}/\{1 + (U/2S)/k_B\eta_1\} \tag{9}$$

where

η_{10} = the loss factor of the bare panel
η_1 = the loss factor of panel + core

both of which are determined empirically.

From Figure 8(a) the frequency dependence of the effective panel loss factor η can be expressed by the empirical relationship

$$\log \eta = \log A - B \log f \tag{10}$$

where the coefficients A and B are determined from the respective charts. The IL of the foam core without septum may be calculated from Equations (9) and (10) using

$$\text{IL} = 10 \log (P_{10}/P_1)$$

Behaviour predicted in this way for a range of effective panel loss factors is illustrated in Figure 9. The values used embrace the range measured empirically and were 0.03, 0.02 and 0.01 at 100 Hz. It is seen that (a) at low acoustic frequencies the IL of the core would be expected to lie in the range 4–8 dB, (b) the IL should decrease modestly with increasing frequency and (c) the incremental enhancement of IL decreases with increasing loss factor, as expected from Equation (6). The decrease in IL with increasing frequency is a result of the fact that

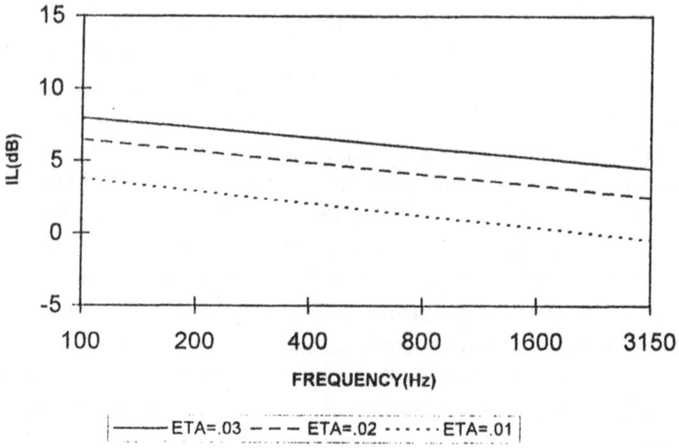

Figure 9. Insertion Loss for vibration generated noise of foam core alone predicted by Equation (9) for three levels of effective panel loss factor.

the empirical difference between the loss factors for bare panel and panel + core decreases as the frequency increases, Figure 8(a). Measured IL behaviour is presented in Section 6.1.

4.1.3 Vibration Insertion Loss–Approximate Model

The predictions of a simple model representation for vibration IL behaviour have been described previously [7]. The model developed below uses a different, more consistent, approach, and incorporates the frequency dependence of the effective panel loss factor. The previous model assumed the loss factor is constant.

The acoustic power P radiated by a panel of area S is related to the average root mean square velocity of the radiating surface $<v^2>$ by [16e]

$$P = \varrho_0 c_0 S \sigma <v^2> \tag{11}$$

where σ is the radiation efficiency.

Assuming the radiation efficiency of the septum layer is equal to that of the bare panel the structure borne noise insertion loss of the composite at a given frequency may be expressed as

$$\mathrm{IL} = 10 \log |P_{10}/P_2| = 10 \log |<v_{10}^2>/<v_2^2>| \tag{12}$$

where P_{10} is the power radiated by the bare metal panel and P_2 the power radiated by the septum with composite in place. $<v_{10}>$ is the average mean square velocity of the bare panel excited by harmonic force F and $<v_2>$ the average mean square velocity of the septum in a panel + core + septum system with the panel excited by the same force.

For a harmonic force excitation F the velocity of a structure $v = j\omega u$ where u is the displacement so the insertion loss may be calculated from

$$\text{IL} = 10 \log |u_{10}/u_{2l}|^2 \tag{13}$$

As an approximation the system may be represented by the lumped model shown in Figure 10.

The equations of motion are

$$(-\omega^2 m_2 + K_2^*)u_2 = K_2^* u_1 \tag{14}$$

$$(-\omega^2 m_1 + K_{12}^*)u_1 - K_2^* u_2 = F \tag{15}$$

where $K_1^* = K_1'(1 + j\eta_1)$, $K_2^* = K_2'(1 + j\eta_2)$ and $K_{12}^* = K_1^* + K_2^*$. m_1 is the surface mass of the panel, K_1' effective in-phase stiffness, calculated from $f_{11} = (1/2\pi)\sqrt{K_1'/m_1}$ where the first mode frequency ~11 Hz, and η_1 is the empirical loss factor of the panel with composite in place, Equation (10). The effective core modulus and loss factor are taken, by the law of mixtures, as the frame (cellular matrix) in parallel

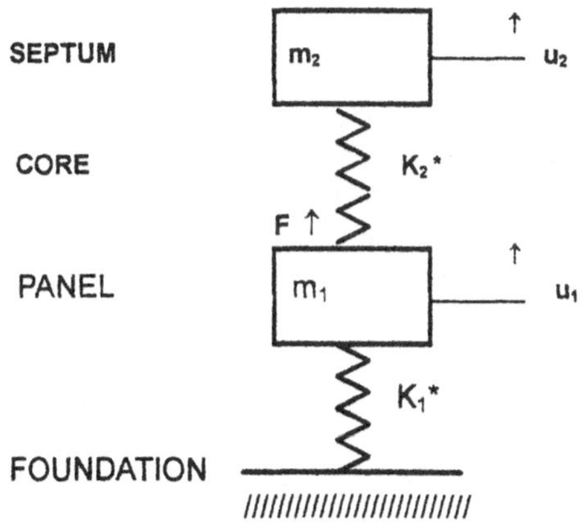

Figure 10. Lumped parameter model for vibration generated noise.

with the air enclosed by the frame which is the same procedure as used for airborne noise, [Equations (26) and (28)]. So the effective core stiffness K_2' and loss factor η_2 are $K_2' = (E_f' + K_{air})/h$ where E_f' is the storage modulus and h the thickness of the core, and $\eta_2 = \eta_f$ where η_f is the loss factor of the core.

From Equation (14)

$$u_1 = \{(-\omega^2 m_2/K_2^*) + 1\}u_2$$

substituting for u_1 in Equation (15) gives

$$u_2 = (F/K_{12}^*)/\{(1 - \omega^2 m_1/K_{12}^*)(1 - \omega^2 m_2/K_2^*) - K_2^*/K_{12}^*)\} \quad (16)$$

For the bare metal panel take Equation (13) and set $m_2 = 0$ and $K_2^* = 0$. This gives

$$u_{10} = (F/K_{10}^*)/\{1 - \omega^2 m_1/K_{10}^*\} \quad (17)$$

where $K_{10}^* = K_{10}(1 + j\eta_{10})$ and $K_{10} = K_1$, the empirical stiffness of the bare metal panel, which is the same as that with the composite in place, and η_{10} = the empirical loss factor of the bare panel, which is much smaller than with the composite in place. Equations (16) and (17) can be substituted into (13) and rationalised to give real, Re, and imaginary, Im, terms. The vibration IL of the composite is then given by

$$\mathrm{IL} = |u_{10}/u_2|^2 = 10 \log\{\mathrm{Re}^2 + \mathrm{Im}^2\} \quad (18)$$

Behaviour predicted by Equations (16), (17) and (18) for typical system parameters is illustrated in Figure 11.

The lumped vibration IL model does not account for wave propagation through the core which influences the behaviour at frequencies above resonance so the results only provide an insight to the effect of design variables at low acoustic frequencies. They are not expected to provide quantitative predictions over the entire frequency range of interest. The range of carpet system parameters studied is given in Table 3.

We see from Figure 11(a) that at ~100 Hz the vibration IL of the composite is expected to be positive and then decrease to a broad minimum at the resonance frequency f_R. Above resonance the IL increases with increasing frequency. Decreasing the area mass m_2 of the septum decreases the IL level and shifts f_R to higher frequencies, as would be

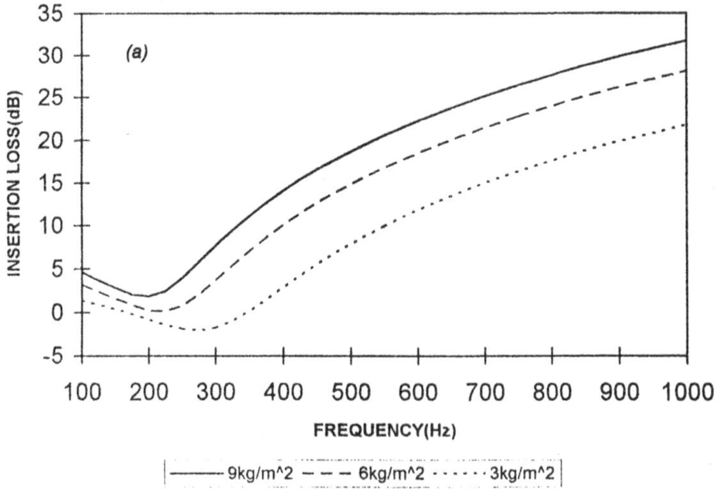

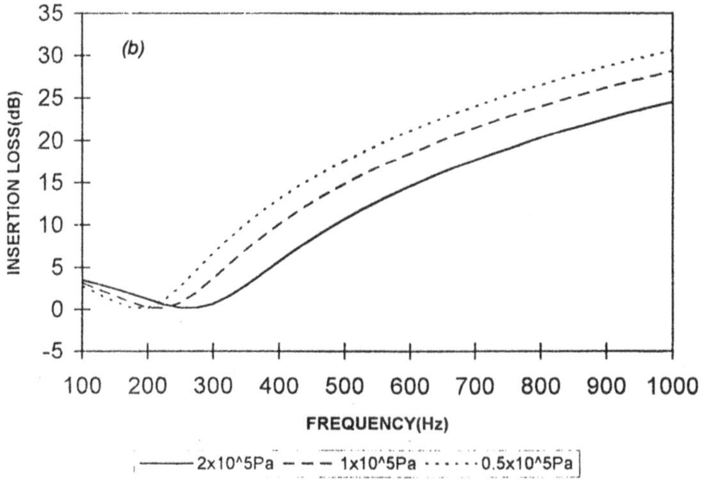

Figure 11. *Predicted Insertion Loss behaviour for vibration excitation. (a) Effect of surface mass of septum; (b) Effect foam core thickness; (c) Effect foam modulus; (d) Effect foam loss factor.*

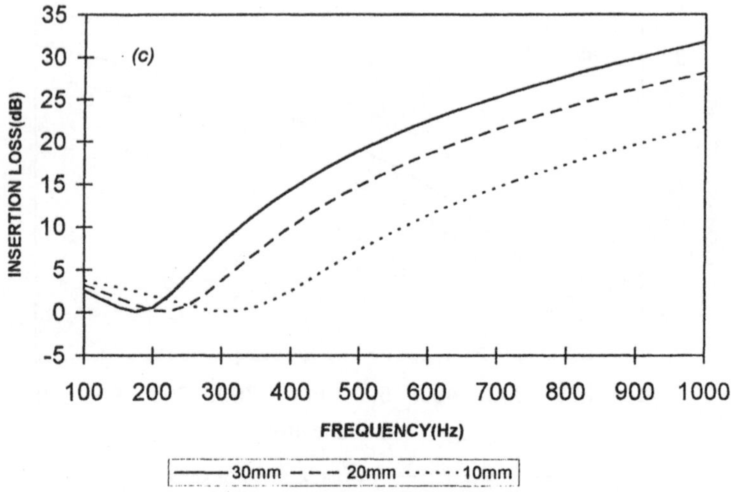

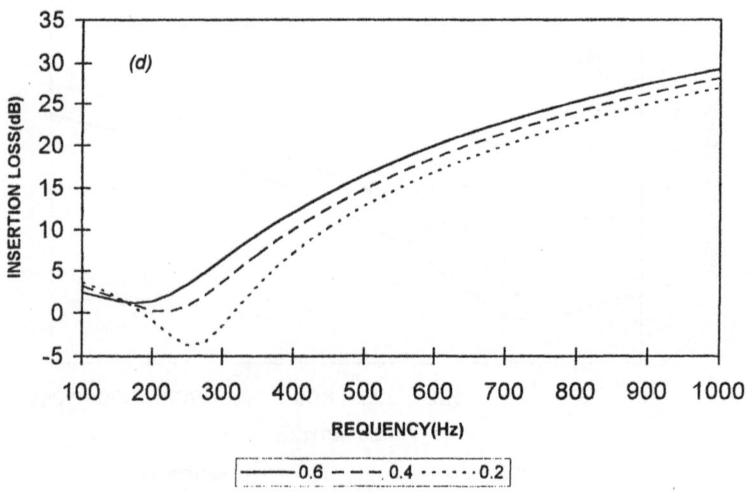

Figure 11 (continued). *Predicted Insertion Loss behaviour for vibration excitation. (a) Effect of surface mass of septum; (b) Effect foam core thickness; (c) Effect foam modulus; (d) Effect foam loss factor.*

Table 3. Range of variables used in the exploration of the predicted vibration IL behaviour.

Figure 11	m_2 (kg/m^2)	E_f' (10^5 Pa)	h (mm)	η_f
(a)	9,6,3	1.0	20	0.4
(b)	6	2.0, 1.0, 0.5	20	0.4
(c)	6	1.0	30,20,10	0.4
(d)	6	1.0	20	0.6, 0.4, 0.2

anticipated. Decreasing the modulus of the foam core, Figure 11(b) should shift the resonance to higher frequencies, again as expected, but is predicted to have little effect on the IL value at resonance. Similar behaviour is predicted for the effect of decreasing the foam core thickness, Figure 11(c). The effect of increasing the foam loss factor, Figure 11(d), is predicted to have the largest effect around the resonance frequency, again as expected from basic vibration theory.

4.2 Airborne Noise

Although the distributed (or transmission line) model of Beranek and Work [17] does not account for all aspects of a practical situation, such as coupling between airborne and structure borne noise inputs, the effect of panel boundaries which create the reverberant field, and the effects of panel damping, it has been widely used as a means for evaluating on a comparative basis the effect of system design variables [23–25]. This has been the approach adopted in our work.

4.2.1 Model Development

The three layer system consisting of panel + core + septum immersed in an air medium of infinite extent is shown in Figure 12. At each interface part of the incident sound energy is transmitted and part reflected the ratio of the two depending on characteristic impedance of the media on either side of the interface. From Equation (1) the Transmission Loss of the system can be expressed in terms of the ratio of the pressure amplitudes on the incident and transmitting sides, which, according to Beranek and Work [17] is given by the product of the pressure ratios at each interface in the system, i.e.

$$p_i/p_t = (p_i/p_1)(p_1/p_2)(p_2/p_t) \tag{19}$$

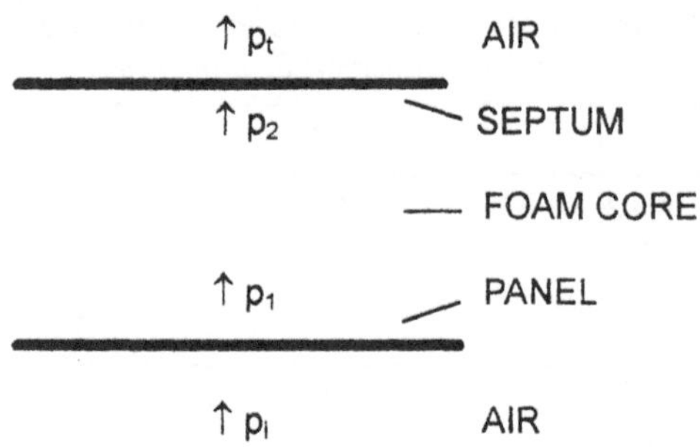

Figure 12. *The acoustic transmission line concept for a carpet composite.*

These pressure ratios can be written in terms of the acoustic impedances on the two sides of the respective interfaces. Assuming the panel and septum have no stiffness, and hence no damping, their impedances can be represented by $j\omega m$ where m is the respective area masses, m_1 and m_2. Thus the impedance Z_i seen from the incident side is given by

$$Z_i = Z_1 + j\omega m_1$$

where Z_1 is a complex function of the parameters which govern the propagation of the wave field through the elastic porous core (modulus, density, flow resistivity) which is terminated by the impervious septum.

Thus at the first interface the pressure ratio is given by

$$p_i/p_1 = (Z_1 + j\omega m_1)/Z_1 = 1 + j\omega m_1 Z_1 \tag{20}$$

The development of the model could proceed in the same way but the simplest approach is to adapt the generalised equations given by Beranek and Work to suit the needs of our system. This is done by taking model IV (p. 422 in the reference cited) and setting the air gap thickness to $l = 0$. This yields

$$p_i/p_t = X(1) \cosh (bh) + X(2) \sinh (bh) \tag{21}$$

where b is the wave propagation constant in the core and h the core thickness. The coefficients $X(1)$ and $X(2)$ are given by

$$X(1) = 1 + j\omega(m_1 + m_2)/Z_{air} \tag{22}$$

$$X(2) = Z^*/Z_{air} - \omega^2 m_1 m_2/Z^* Z_{air} + j\omega m_1/Z^* \qquad (23)$$

where $Z_{air} = \varrho_0 c_0$ is the acoustic impedance of free air and Z^* the complex acoustic impedance of the core material.

For the core material the propagation constant b and acoustic impedance Z^* are given by the relationships [26]

$$b = j\omega(\varrho^*/K_T^*)^{1/2} \qquad (24)$$

$$Z^* = (\varrho^* K_T^*/Y)^{1/2} \qquad (25)$$

where Y is the porosity, K_T^* the effective complex bulk modulus and ϱ^* the effective density. With the system configuration used in the test rig in which the top, bottom and edges of the porous core, although not bonded, are effectively sealed we assume that there is no net flow of air into or out of the core. We take the effective modulus as being the modulus of the frame in parallel with the air enclosed by the frame, i.e.

$$K_T^* = K_f^* + K_{air}^* \qquad (26)$$

where, in the frequency range of interest the in-phase bulk storage modulus of air, K'_{air}, ranges from its isothermal value to its adiabatic value, ~ 1.1 to 1.4×10^5 Pa. In this application the loss factor of the air enclosed by the framer is unimportant. The value K'_{air} adopted is important since it is of the same order as that of a typical core frame and the choice of the value leads to uncertainty in the predicted response, particularly the resonance frequency, f_R.

For an isotropic homogeneous material the bulk modulus K is related to the Young's modulus E by $K = E/3(1-2\nu)$ where ν is the Poisson's ratio. For a wide density range of open cell plastic foams Gibson and Ashby [27] showed empirically that the average value of ν is ~ 0.33. For open cell elastomeric foams Gent and Thomas [28] predicted theoretically that $\nu = 0.25$. Taking the Gibson-Ashby value the bulk modulus of the frame is shown to be $K_f \sim E_f$. In our model representation we take the elastic modulus of the core as being complex and given by

$$E_f^* = E_f'(1 + j\eta_f) \qquad (27)$$

where E_f' and ν_f are measured in a uniaxial vibration transmissibility test (i.e. the BMW test). This is a resonance method in which the test frequency is controlled by the material and usually lies in the range 50–250 Hz.

The dissipative nature of fluid flow through a porous medium is often taken into account by defining an effective fluid density. Several approaches have been described in the literature ranging from highly complex to relatively simple [18,26,29]. One simple procedure is to take the total density of the medium through which the wave propagates as

$$\varrho^* = \varrho_f + \varrho_0(1 - jR_1\, Y/\omega\varrho_0 k_s) \tag{28}$$

where ϱ_f is the foam density, R_1 the static airflow resistivity [30], Y the porosity, and k_s the structure factor of the core foam. The Kosten and Zwikker structure factor k_s [26] is related to the tortuosity of the foam, [30], and for the flexible and viscoelastic foams studied in our work the value was ~1.0.

The transmission loss of the panel with carpet composite in place $R(2)$ is calculated by combining Equations (21), (22) and (24). The situation for the panel alone without the composite in place $R(1)$ is simulated by setting the septum area mass $m_2 = 0$ and the core thickness $h = 0$. This gives cosh $(bh) = 1$ and sinh $(bh) = 0$ and

$$p_i/p_t = 1 + j\omega m_1/Z_{air} \tag{29}$$

so

$$R(1)\,(\text{dB}) = 10 \log \{1 + (\omega m_1/Z_{air})^2\} \tag{30}$$

which differs from the limp panel mass law, Equation (2), in terms of the factor of 2 in the second term in the {}. The explanation for this difference is that Equation (30) gives the ratio of the noise on the receiving and transmitting sides with the panel in place whereas Equation (2) gives the ratio of noise levels at the two points in space, i.e., the incident side and receiving side, before and after the panel has been put in place. The insertion of the panel affects the noise field on the incident side and according to [17] the two situations are different.

The predicted insertion loss of the composite is given by

$$\text{IL} = R(2) - R(1) \tag{31}$$

4.2.2 Approximate Solution

At low acoustic frequencies we can make the approximations cosh $(bh) = 1$ and sinh $(bh) = bh$. Substituting into Equation (21) gives

$$p_i/p_t = \{1 - \omega^2 m_1 h/K_T^*\} + j\{\omega(m_1 + m_2 + \varrho^* h)/Z_{air} - \omega^3 h m_1 m_2/K_T^* Z_{air}\} \tag{32}$$

Rationalizing yields real and imaginary parts and the resonance frequency is predicted as

$$\omega_R = \sqrt{(K_T'/h)\{m_1 + m_2 + (\varrho_0 + \varrho_f)h\}/m_1 m_2} \tag{33}$$

where K_T' is the total in-phase bulk modulus of the core. As an approximation the density term is ignored.

This approach produces results similar to those given above by the lumped parameter model for the vibration Insertion Loss behaviour. These are illustrated using the system parameters given in Table 3 in Figure 13. At 100 Hz the airborne noise IL value is predicted to be a little higher (2–3 dB) than that for structure borne noise but the resonance trough is deeper and the frequency of resonance higher. With both models the predicted IL values above the resonance trough are larger than would be expected because wave propagation effects are not accounted for in these approximations. However they are clearly seen when the cosh (bh) and sinh (bh) terms are retained in the exact solution for airborne IL.

4.2.3 Exact Solution

Predictions for the system parameters given in Table 3 are illustrated in Figure 14. Below and around the resonance frequency the predictions of the two models are much the same. However above resonance the exact solution is different from the approximate solution, the IL values rising less steeply and troughs appear at frequencies above 1 kHz. These are associated with the cosh and sinh terms in Equation (21) and are due to the effect of wave propogation through the core. The predictions of this model are compared with measured behaviour in Section 7 below.

5. TEST METHODS

Procedures for the evaluation of noise control composites are divided into two categories, in-car either on-road or laboratory based [4,5,12,31], and laboratory tests. Some laboratory based procedures for measuring IL behaviour and the characterisation of the materials used

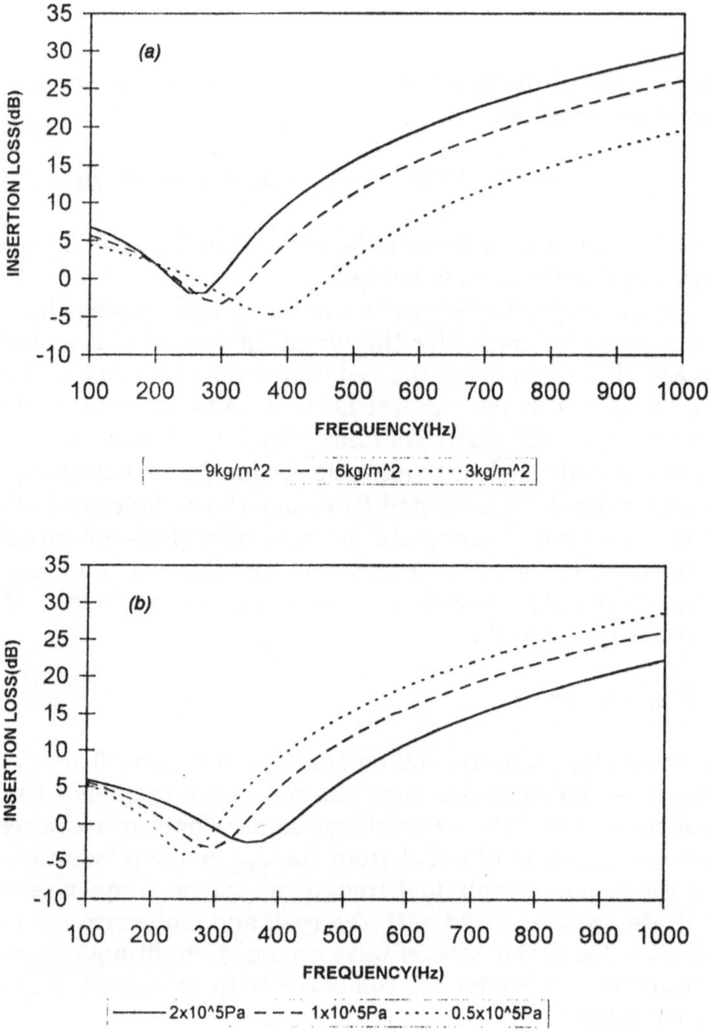

Figure 13. *Insertion Loss behaviour for airborne noise predicted by the approximate model. (a) Effect of septum surface mass; (b) Effect of foam modulus.*

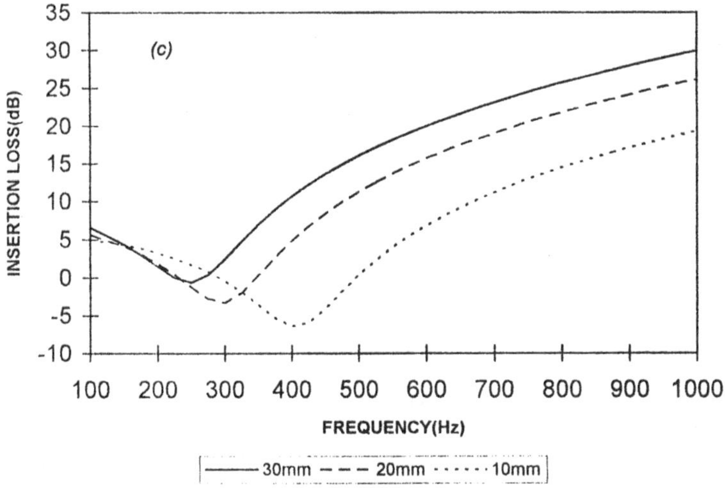

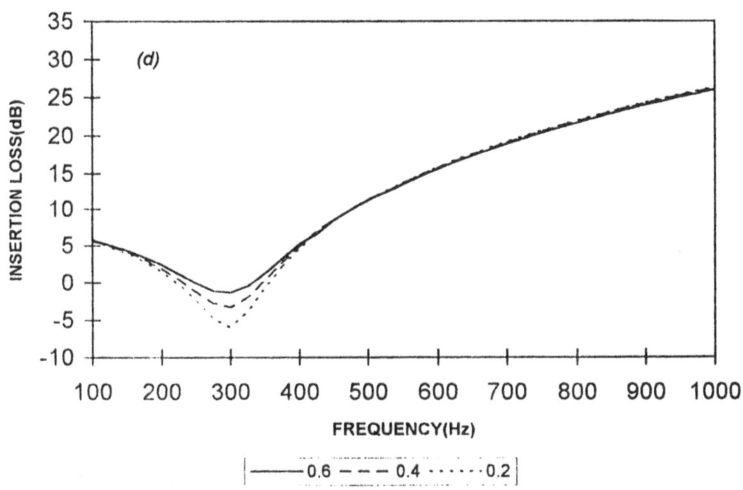

Figure 13 (continued). *Insertion Loss behaviour for airborne noise predicted by the approximate model. (c) Effect of foam core thickness; (d) Effect of foam loss factor.*

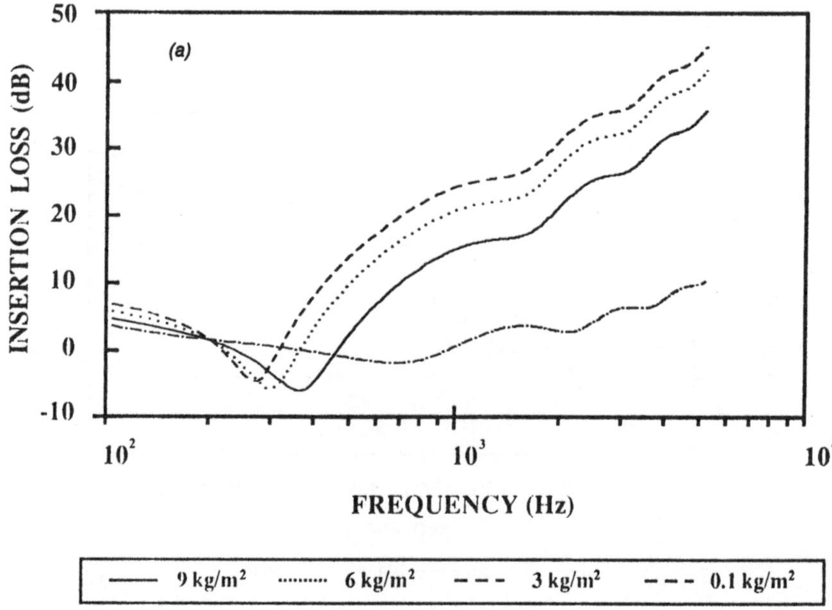

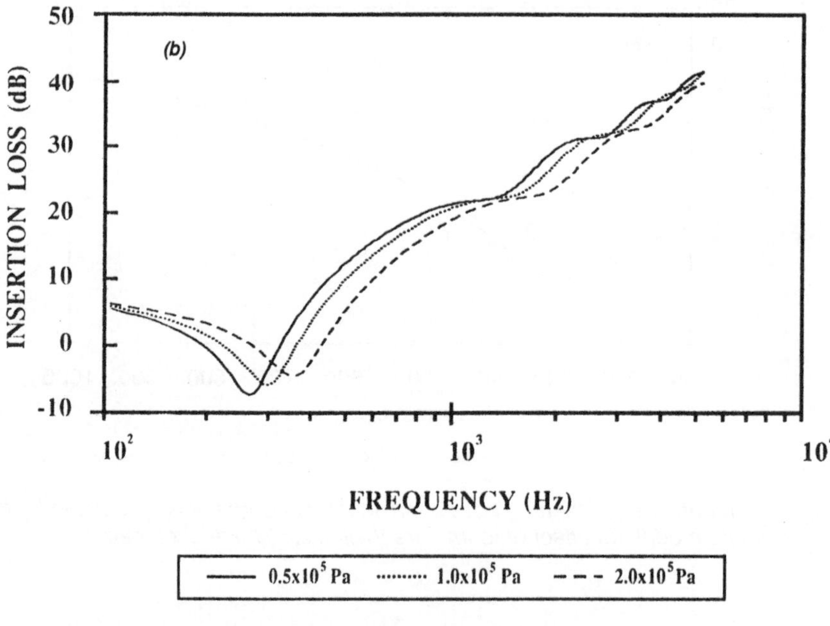

Figure 14. *Insertion Loss behaviour for airborne noise predicted by the exact solution. (a) Effect of septum surface mass; (b) Effect of foam modulus.*

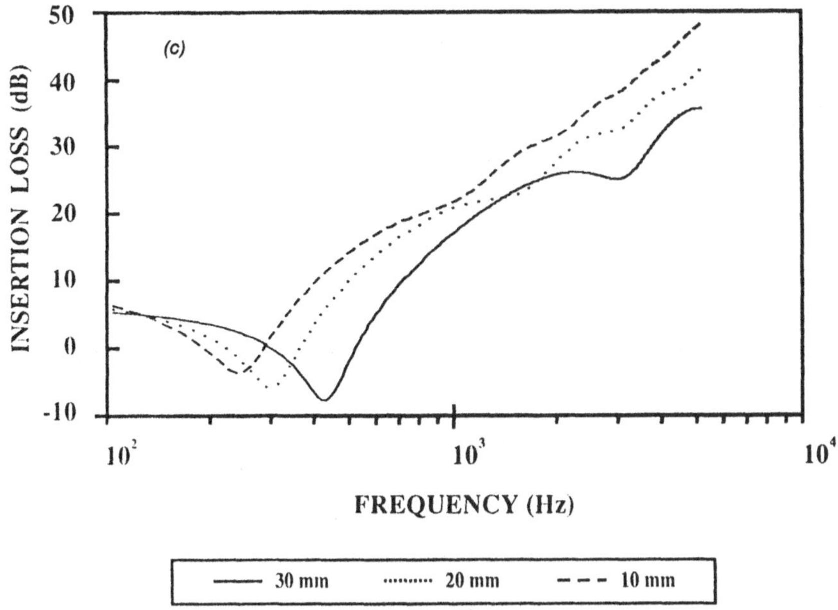

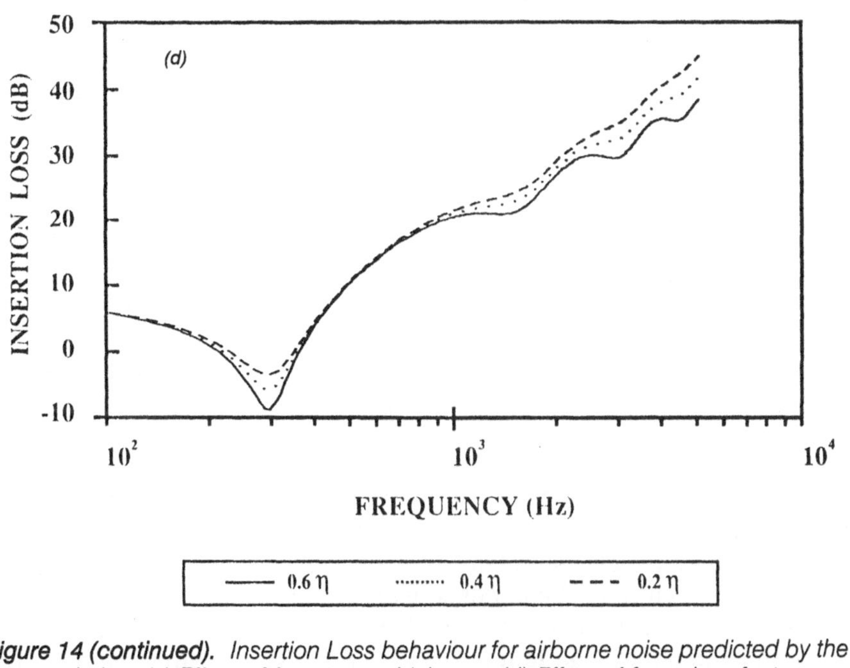

Figure 14 (continued). *Insertion Loss behaviour for airborne noise predicted by the exact solution. (c) Effect of foam core thickness; (d) Effect of foam loss factor.*

in the systems are described in Reference [32]. Although in-car testing will feature somewhere in a development programme it is time consuming and it is difficult to separate the effect of floorpan/bulkhead from the "other" sources. Furthermore the complex geometry of the real situation does not lend itself to theoretical analysis so in-car measurements are not entirely suited to fundamental studies of system behaviour. The alternative is to use laboratory based procedures which have "ideal" geometry. They fall into two categories, large scale e.g., the standard transmission loss suite [35], and small scale. The technical disadvantages of the standard transmission suite are that the test structure is mounted vertically so in the present situation the foam core would have to be bonded to the metal panel and there is no facility for vibration excitation. The current trend in the automotive industry is to employ lay-on carpet systems so the vertical configuration of the standard suite is not entirely appropriate. Horizontal, or 45 degree suites are available but these are rare, and in the case of the 45 degree panel the results give rise to uncertainties as to the effect of gravitational forces on the interaction between the foam core and the panel.

A number of small scale systems have been described e.g., the Apamat II apparatus [33], and the zero impedance apparatus [34] used originally by Beranek and coworkers in their studies of noise transmission through multi-element aircraft panels. The Apamat II apparatus consists of two parts—a sending chamber and a receiving chamber which is shaped to avoid the development of standing waves and the reverberation time adjusted to correspond to the conditions that obtain within a vehicle interior. Structure borne excitation is provided by means of 8 mm steel balls projected onto the underside of the steel panel and airborne is from loudspeaker broadband random noise. The SPL is measured on the receiving side without and with the composite in place and the SPL IL determined from the difference in these values. This procedure does not lend itself to fundamental studies of system behaviour which require well defined vibration and acoustic inputs.

The zero impedance apparatus has been adopted and adapted by the automotive industry, e.g., GM and Ford. In these adaptions the test panel is horizontal and the sending side is replaced by a rectangular double walled chamber with sand infill to reduce wall vibration. An array of fifteen speakers supplemented by nine high frequency tweeters mounted very close to the panel is used as the sound source. The chamber, which is lined with acoustic absorbing material, is about 0.6 m deep and the panel side length about 0.8 m. The arrangement on the receiving side varies. In some cases a rectangular anechoic receiving chamber about 1 m high is employed whilst in others the system radiates into an anechoic room. The sound intensity level is measured on the receiving side and the composite characterised in terms of the level difference without and with the composite in place. As described in the literature this apparatus has no facility for vibration excitation and in our initial work on test rig development it was found that care has to be exercised to avoid leaks and flanking which, in our experience, can affect high frequency results i.e. at $f > 2$ kHz.

The Beranek zero impedance apparatus has an important practical limitation for our purposes, i.e. the facility to excite the panel by means of mechanical vibration. As a compromise between the acoustic requirements of the sending chamber and receiving environment, the ability to vibrate the panel, ease of use and cost the test rig shown in Figure 15 was developed. To avoid difficulties associated with flanking the sending chamber was of brick/mortar construction and mounted on a reinforced concrete base which was vibration isolated from the concrete laboratory floor. The plinth was also of reinforced concrete and contained a 0.87 m × 0.87 m aperture. The airborne noise field was created by a 100 W low frequency drive unit (50 Hz–3.5 kHz) augmented by two piezo tweeters (2.2 kHz–33 kHz). A 1.0 mm thick steel panel was placed over the aperture, with flexible neoprene sealing

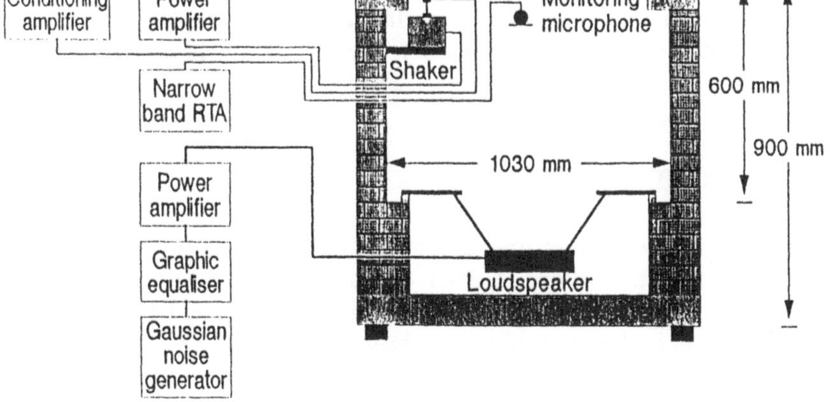

Figure 15. The test rig.

strip around the boundaries, and secured by means of a 4 mm thick angle iron framework and quick release toggle clamps. Broad band excitation of speakers and shaker was achieved via a sine/random generator, graphic equaliser and appropriate power amplifiers. The shaker was loosely connected to the panel by means of an elastomeric coupler situated ~0.15 m × 0.15 m from the panel boundary. This location was chosen to ensure that low frequency panel modes were not strongly excited. The panel radiated into a large, open, laboratory with reverberation time ~2 sec and the noise level measured by sweeping a two microphone intensity probe over the entire surface of the test piece about 100 mm from the surface.

Since the sound field on either side of the panel was not diffuse transmission loss measurements were not possible. However the arrangement could be used for measuring the IL of composites since measurements were only made on the receiving side. The internal noise spectrum was monitored using real time analysis to ensure that the level in the sending chamber was approximately constant across the frequency range of interest and the same from day to day. The values were not used for analysis purposes. Reproducible data could only be obtained when the laboratory was quiet and under these conditions tests on the same systems over a period of several months showed that for the frequency range of prime interest, 100 Hz–3.15 kHz, the reproducibility in airborne IL values was about ±1 dB.

The experimental IL data presented below are given as 1/3 octave band levels and were measured using non-bonded systems (panel/core and core/septum) with PU foam cores of density ϱ_f ~80 kg/m³ and modulus values E_f' in the range ~0.4–2.0 × 10^5 Pa. Unless otherwise stated the core thickness was 20 mm and the septum was commercial grade EVA automotive carpet heavy layer without carpet. The effect of tufted carpet is considered separately. The foams are identified as A, B, C, . . . O.

6. TEST RESULTS–VIBRATION

Vibration IL behaviour has not been studied as extensively as airborne behaviour and there is little or no experimental evidence from other workers to corroborate the results reported here. In our case the reasons for this is that the vibration measurements were less reproducible than those for airborne noise, the lumped model does not account for wave propagation effects, which are significant at $f > f_R$ and there was less confidence in the credibility of measured behaviour. However, a review of available data provides an insight to the situation. The data shown are for three design variables m_2, E_f' and η_f typical of those under development or used in-car. Values of parameters explored are given in Table 4.

Before exploring the effect of these variables we first examine the IL of foam cores alone (no septum).

6.1 Foam Core

The vibration IL behaviour of several nonbonded cores, having the same thickness but different physical properties, is shown in Figure 16. It is seen that there are differences in behaviour between the foams and that for an individual foam there was considerable scatter in results; larger than those observed with incident airborne noise (Figure 21). Examination of the data overall reveals some degree of consistency. The main observations are:

1. The mean IL value over the frequency range studied is ~7 dB which is the same as that predicted for low acoustic frequencies in Section 4.1.2 and is significantly larger than that for the airborne noise IL of the core alone below 1 kHz.
2. The mean behaviour does not fall off with increasing frequency as predicted. This is because the empirical panel damping behaviour used in the model prediction indicates that the loss factor for panel + core falls off more rapidly than that for the bare panel.
3. There are two troughs in the response (a) in the 1/3 octave bands centred at ~125–250 Hz and (b) in the bands centred at ~630–1000 Hz. The origin of the first is not known but is observed to a greater or lesser degree in the vibration IL behaviour of core + septum systems, Figures 17–19. This trough is not seen in the airborne IL response. It is speculated that the higher frequency trough is associated with the resonance vibration of the nonbonded foam core on the panel.

Table 4. System parameters for measured vibration IL data presented in Figures 17–19.

Figure	m_2 (kg/m²)	E_f' (10^5 Pa)	η_f
17	3.7, 5.9, 8.7	~0.5	~0.46
18	5.9	1.9, 1.2, 0.5	~0.41–0.46
19	5.9	~0.4–0.6	0.91, 0.46, 0.16

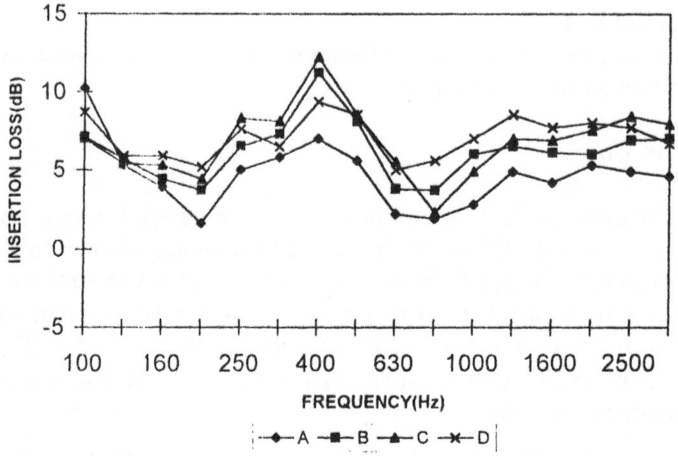

Figure 16. Measured vibration Insertion Loss behaviour for PU foam cores without septum.

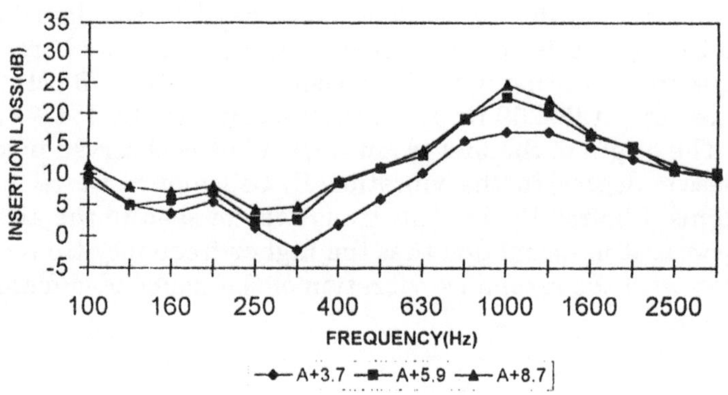

Figure 17. Measured vibration Insertion Loss behaviour: Effect of septum surface mass m_2.

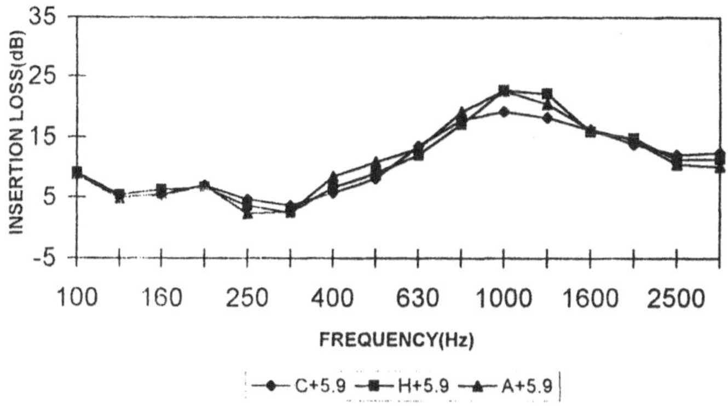

Figure 18. Measured vibration Insertion Loss behaviour: Effect of foam modulus E_f'.

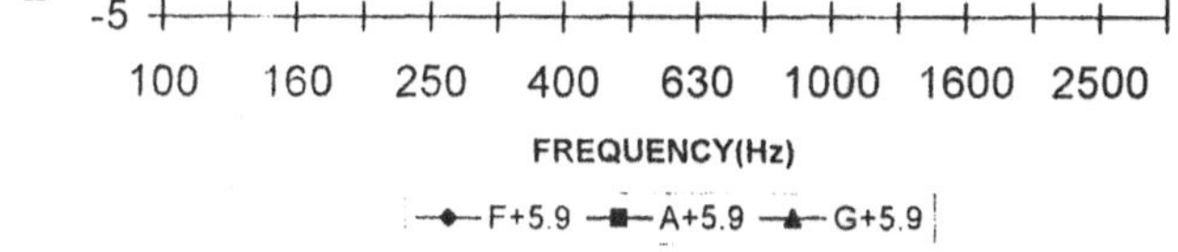

Figure 19. Measured vibration Insertion Loss behaviour: Effect of foam loss factor η_f.

6.2 Septum

Figure 17 shows typical results for the influence of septum mass, m_2, on the IL behaviour. At low acoustic frequencies the IL value is positive and falls to a broad minimum as the frequency increases to the resonance trough. At 100 Hz the IL value is ~10 dB, which is ~3–5 dB larger than that measured for airborne noise Figure 22 and at this frequency is ~5 dB larger than that predicted by the model, Figure 11(a). However the value of the resonance frequency, f_R, is in reasonable quantitative agreement with that predicted and the data shows it to increase with increasing value of m_2, albeit modestly, and the depth of the trough to decrease, as anticipated. Above f_R, where the lumped model does not apply the IL increases strongly to a maximum and then falls into another trough somewhere above the 1/3 octave band centered at 3.15 kHz. The origin of this high frequency trough is not known. It is observed in the airborne noise IL response and is not thought to be due to wave propagation through the core since this can be predicted in the case of incident airborne noise. It might be a manifestation of coincidence in the septum or the result of edge effects.

6.3 Foam Modulus

Figure 18 compares the response of three composites with different core modulus E_f' but with the same EVA septum, $m_2 = 5.9$ kg/m². The modulus values are given in Table 3. We see that increasing E_f' in the range 0.5–1.9 × 10^5 Pa shifts the values of f_R from the 1/3 octave band centered at 250 Hz to 315 Hz, which is in general agreement with the behaviour predicted by the model. The small decrease in the depth of the trough with increasing f_R is similar to that observed with airborne noise (Figure 25) but is not predicted by the model. It is noted that for the same systems, at resonance the vibration IL is ~3 dB larger than the airborne IL.

6.4 Foam Loss Factor

It has been difficult to establish unequivocally the influence of the loss factor of the foam because of the difficulty in holding other variables constant; such as foam modulus, density, flow resistivity. The model indicates that this parameter should primarily affect the behaviour at and around the resonance frequency and that for the range studied there should be significant differences. Typical data are illustrated in Figure 19 from which it is seen that in the range $\eta_f \sim 0.2$–0.9

the depth of the trough decreased by ~2 dB which is less than expected. Similar observations are made with airborne noise and in this case it can be attributed to the fact that the model does not account for panel damping. However in the case of vibration excitation this is incorporated by employing empirical values of effective panel loss factor, i.e. panel + core. The insensitivity of measured IL behaviour at resonance to variations in foam loss factor cannot be explained at this time.

6.5 Carpet

The model assumes that the only role the fibrous carpet plays is that of adding a relatively small mass to the septum, m_2. This assumption was tested empirically by placing a carpet on top of the EVA septum in a number of composite systems and measuring the intensity on the receiving side without and with the carpet in place. The results are shown in Figure 20 from which it is seen from the overall behaviour below ~400 Hz the IL value is ~1 dB. This behaviour is in quantitative agreement with that measured for airborne noise (Figure 28) and is consistent with adding a small mass to m_2. In the frequency range ~800–1600 Hz the response passes through a trough and becomes negative. This trough is observed in the airborne noise response although it is not as intense. As with the higher frequency trough observed with core alone, Figure 16, it is attributed to a vibration resonance, in this case the carpet on the septum.

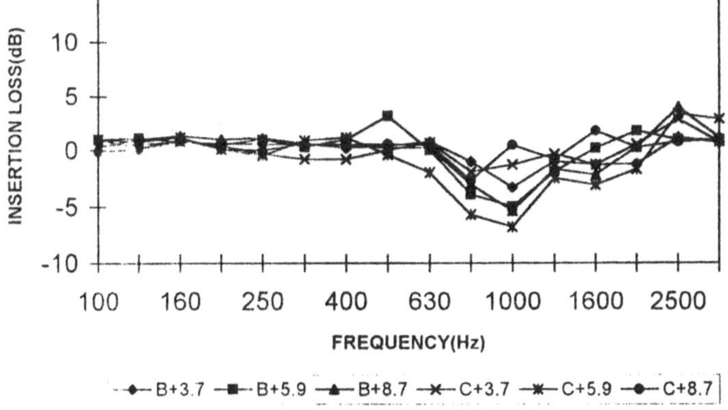

Figure 20. Measured vibration Insertion Loss of a carpet placed on different core + septum systems.

7. TEST RESULTS–AIRBORNE

The effect of system design parameters on the airborne noise Insertion Loss behaviour has been studied in some detail. As indicated above it was not always possible to explore the behaviour of one parameter whilst holding all others constant because of small variations in the physical properties of the manufactured foam cores. The range in variables for the data discussed below are given in Table 5.

7.1 Foam Core

The airborne IL provided by the foam core alone is illustrated for a number of core materials in Figure 21. The range of material parameters was E_f ~0.6–1.3 × 10^5 Pa and η_f ~0.16–0.9. At the 1/3 octave band centred at 100 Hz the IL value is about 3 dB indicating an attenuation of the transmitted noise. As the frequency increases the IL decreases to ~0 dB at 300 Hz. Above 300 Hz it becomes negative, values ranging from about −1 dB to −4 dB, indicating that the noise transmitted by the system is larger than that transmitted by the bare metal panel. Above the trough the IL value becomes positive and unlike the vibration IL behaviour described above the airborne IL increases almost monotonically with increasing frequency.

For the core alone case a septum mass m_2 = 0.1 kg/m² was used in the exact model to approximate the situation for zero septum mass. The code will not run with m_2 = 0. The predicted behaviour is shown in Figure 14(a) from which it is seen that there is excellent agreement between modelling and measurement, with the model correctly predicting the broad negative IL value in the range ~315–800 Hz.

7.2 Septum

The influence of septum properties on behaviour is the easiest and

Table 5. System variables for the airborne IL results.

Figure	m_2 (kg/m²)	E_f' (10^5 Pa)	h (mm)	η_f
22	3.7, 5.9, 8.7	0.6	20	0.16
23	3.7, 5.9, 8.7	1.3	20	0.42
24	3.2–3.7	1.3	20	0.42
25	5.9	0.47, 1.3, 1.9	20	~0.4
26	5.9	0.6	30, 20, 10	0.16
27	3.7	~0.4–0.5	20	0.9, 0.4, 0.16

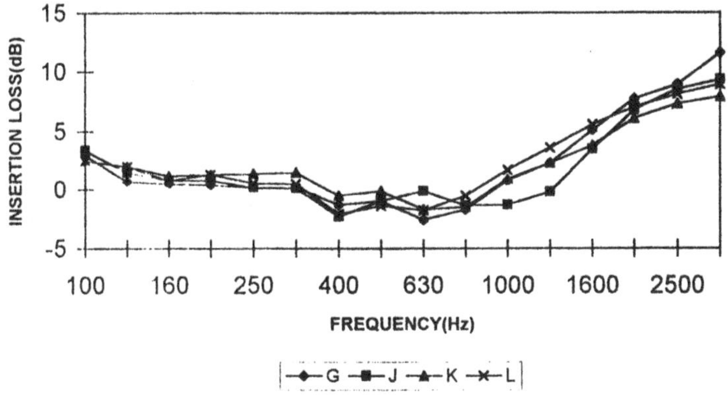

Figure 21. Measured airborne noise Insertion Loss behaviour of PU foam cores without septum.

most reproducible investigation since it does not involve cores of different physical properties; modulus, loss factor, density, flow resistivity, surface adhesion. Studies of this parameter invariably produce consistent results. Typical behaviour for the effect of septum area mass m_2 is illustrated in Figure 22. Increasing the value of m_2 enhances the IL at low acoustic frequencies, shifts the resonance of the panel + core + septum system to lower frequencies and reduces the depth of the trough at resonance. This behaviour is in quite good qualitative and quantita-

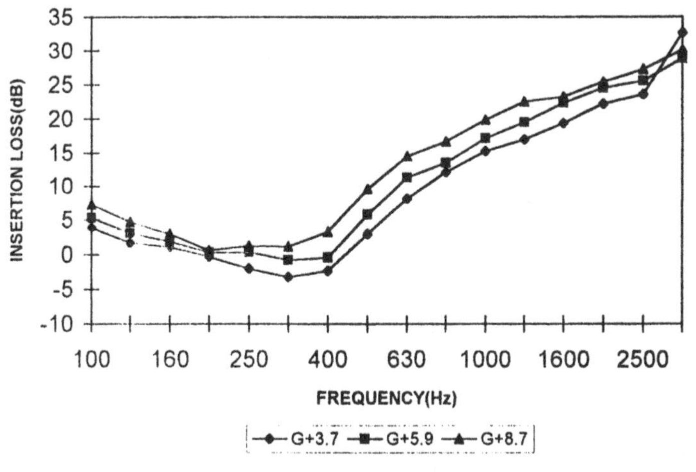

Figure 22. Measured airborne noise Insertion Loss behaviour: Effect of septum surface mass m_2.

tive agreement with the exact solution, Figure 14(a), except at and around resonance where the predicted IL trough is deeper than that measured. The reasons for this are discussed in Section 8.

Investigations have shown that system resonance, at frequency f_R, plays a central role in governing airborne IL behaviour. This is illustrated in Figure 23 in which the responses for three septum area masses are superposed. The resonance frequency for $m_2 = 8.7$ kg/m², 250 Hz, is taken as the reference point and the other curves shifted downwards by a factor a_f. The values of a_f are 1 × 1/3 octave band for $m_2 = 5.7$ kg/m² and 2 × 1/3 octave bands for $m_2 = 3.7$ kg/m². For a limited frequency range above resonance the curves for the different septum masses superpose quite well. Although this is an abitrary procedure it concisely demonstrates the importance of f_R.

The transmission line model of Beranek and Work (Section 3.2) assumes the septum acts as a limp mass i.e., the stiffness and damping of the layer play no role in governing IL behaviour. To test this assumption measurements were made using an industrial rubber matting septum which had low stiffness and damping relative to our standard EVA septum material. Results for two foam cores are given in Figure 24 from which it is seen that in both cases there was little difference between the responses for rubber and EVA septums. The septum masses, m_2, and the properties of the core foams were approximately the same but the system with the higher resonance frequency had core thickness

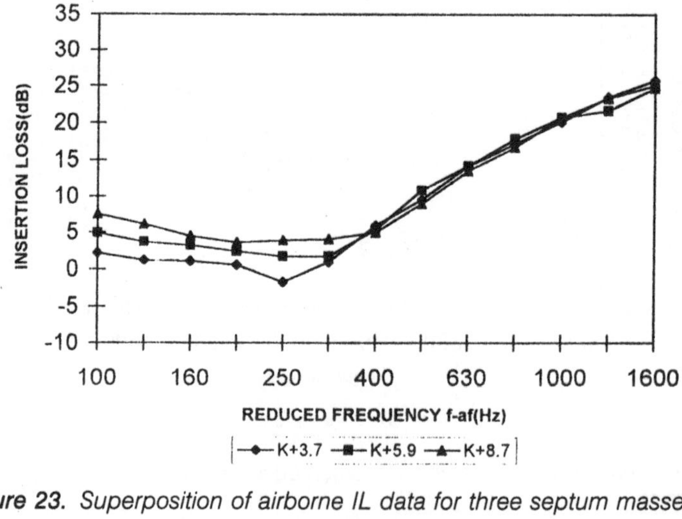

Figure 23. Superposition of airborne IL data for three septum masses, m_2.

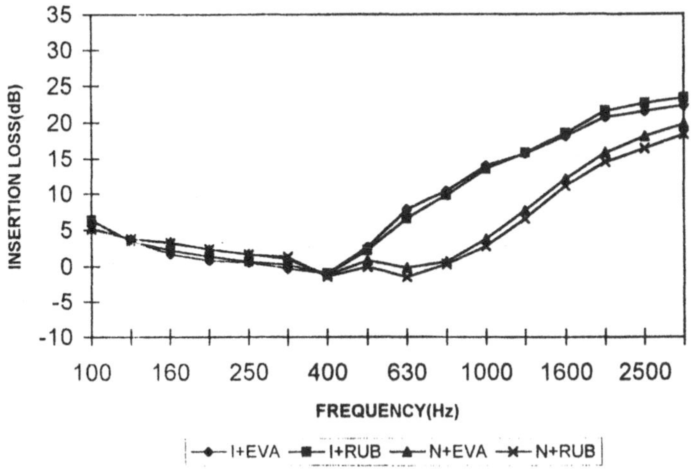

Figure 24. Influence of septum physical properties on the airborne Insertion Loss behaviour for two PU foam cores.

$h = 10$ mm whereas the lower resonance frequency system had core thickness $h = 20$ mm.

7.3 Foam Modulus

Amongst other aspects of the response the modulus of the core foam E_f' is predicted to influence the value of f_R and the depth of the trough at f_R. This turns out to be the case as illustrated in Figure 25. It is seen that increasing the modulus shifts the value of f_R to higher frequencies, about 1/3 octave band, and decreases the depth of the trough by a small amount. The influence of the lower resonance frequency is reflected in the enhanced value of IL above resonance as discussed above. Theoretical predictions of the effects of varying the dynamic modulus of the core foam is shown in Figure 14(b). Again excellent agreement is seen between modelling and measurement apart from an enhanced resonance trough.

7.4 Core Thickness

The foam core thickness h is a major system variable and of considerable practical interest. As with increasing the modulus of the core foam decreasing the thickness of the core would be expected to increase the

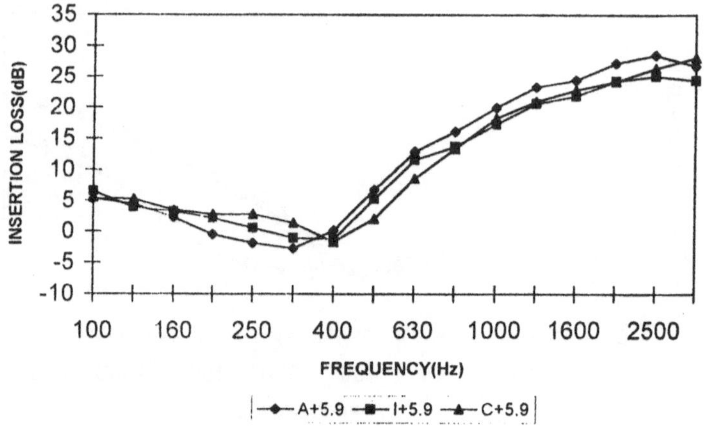

Figure 25. Measured airborne noise Insertion Loss behaviour: Effect foam modulus E_f'.

resonance frequency because of the enhanced stiffness of the isolating layer. The data presented in Figure 26 show that this is the case. However unlike the effect of core modulus decreasing the core thickness causes an increase in the depth of the resonance trough. It is seen that below resonance the IL values increase with decreasing thickness. However above resonance the IL response of the thinner core is significantly lower than that of the thicker cores. All of these aspects of IL

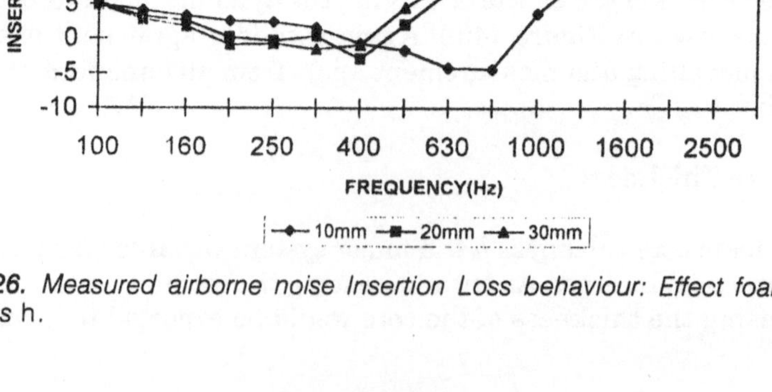

Figure 26. Measured airborne noise Insertion Loss behaviour: Effect foam core thickness h.

behaviour are correctly predicted by the model, as shown in Figure 14(d).

7.5 Foam Damping

Theory predicts that the main effect of foam loss factor η_f is at and around the system resonance frequency and the higher frequency troughs associated with wave propagation through the core, Figure 14(d). Measured behaviour, illustrated in Figure 27, indicates that this is in fact the case but the effect at the primary resonance is much smaller than expected. For the data shown here increasing η_f from ~0.2 to ~0.9 resulted in a decrease in the depth of the airborne IL trough by about 1 dB. The difference in the f_R values for this data is associated with different modulus values (see Table 4). Although similar behaviour has been observed with other core foams the difference illustrated here is within the limits of measurement precision. It is concluded that the effect of foam loss factor is less than that predicted by the transmission line model. The reason for this is discussed in Section 8.

7.6 Carpet

The IL of a tufted carpet placed on the EVA septum for incident airborne noise was measured in the same way as for vibration excitation.

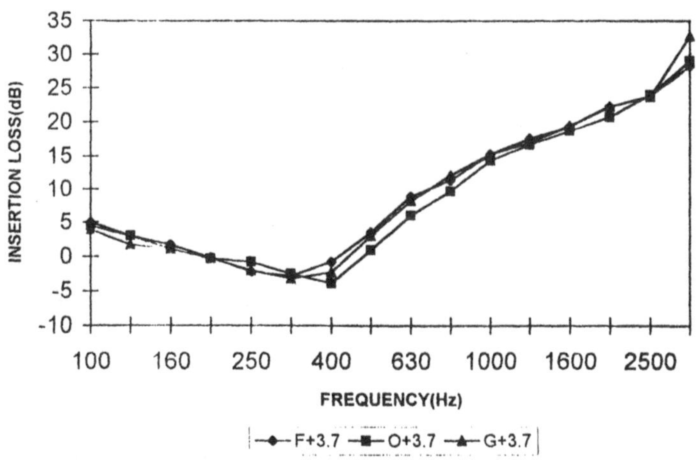

Figure 27. Measured airborne noise Insertion Loss behaviour: Effect foam loss factor η_f.

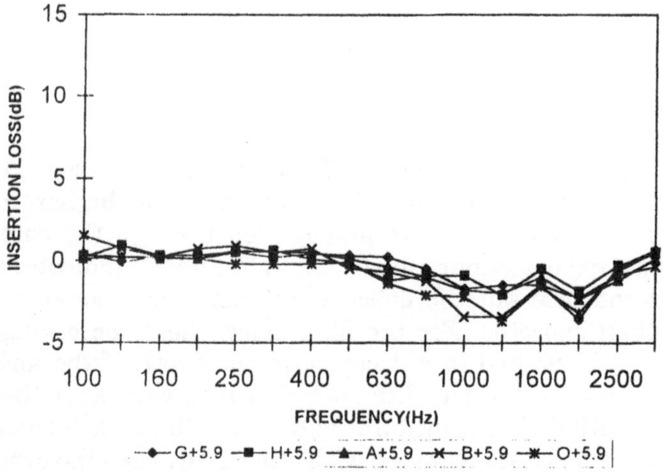

Figure 28. Measured airborne noise Insertion Loss behaviour of a carpet placed on different core foams with 5.9 kg/m² EVA septum.

Some results are given in Figure 28 for a septum mass $m_2 = 5.9$ kg/m². Values of E_f' ranged from ~0.5–2 × 10^5 Pa and η_f from ~0.2–0.5. It is seen that the scatter in the data is less than that for vibration excitation, Figure 20, and the depth of the resonance trough in the frequency range 1–2 kHz is smaller. However, overall the behaviours for the two types of panel excitation are much the same.

8. CONCLUSIONS

The discussion centres around two aspects of the investigations: (a) a comparison between vibration and airborne IL behaviour and (b) the failure of the transmission line modelling approach to accurately predict the depth of the airborne noise IL resonance trough. At this stage it is not worthwhile reconciling the vibration IL model with measurement since it does not account for wave propagation effects.

8.1 Vibration Versus Airborne Behaviour

Laboratory tests and modelling studies have shown that design parameters controlling the IL behaviour of a composite can be divided into two categories: (1) system i.e. heavy layer and core thickness and (2) material i.e. the modulus, air flow resistivity, damping and density of the core foam, although the behaviour is not very sensitive to the last two material parameters. The current work and the small amount of data published in the literature indicate that at low acoustic frequencies the IL value is positive but decreases to a trough at system resonance and then increases as the frequency is increased further. The resonance frequency f_R is controlled primarily by the effective modulus and thickness of the core and the surface mass of the metal panel and septum, m_1 and m_2 respectively. Since the effective core modulus is the modulus of the foam frame in parallel with the air enclosed by the frame and in many practical situations the bulk modulus of the frame is about the same or less than that of air the value of f_R is not strongly dependent on the modulus of the core foam and for a given value of m_2 there is a lower limit to f_R. This condition occurs when $E_f' \sim 0.2 \times 10^5$ Pa and for a septum mass of 5.7 kg/m^2 the lower frequency limit is ~200 Hz. In some cases the IL at resonance may be negative indicating that the composite enhances noise transmission so that the transmitted noise level is larger than that with the bare metal panel.

Measurement and modelling indicate that at low acoustic frequencies, ~100 Hz, for the same system parameters the vibration IL is ~3–5 dB larger than the airborne IL. As the frequency is increased the vibration IL passes through a relatively weak trough before it reaches the primary system resonance. This is not seen in the case of airborne excitation. A similar trough is sometimes observed in IL behaviour derived from in-car noise level measurements but its origin has yet to be explained. At resonance the vibration IL is several dB larger than the airborne IL, as predicted by the approximate models, and in most situations the vibration IL remains positive whereas the airborne IL is often negative at this frequency. Above resonance the vibration IL rolls off at a frequency lower than that for the airborne IL. Thus although the behaviours for structure borne and airborne noise are similar they differ in detail.

8.2 Reconciliation of Measurement with Theory

Overall the predictions of the airborne IL model at frequencies between 100 Hz and 3.15 kHz are in good quantitative agreement with measured behaviour e.g. the IL values at low and high frequencies and the value of the resonance frequency f_R. However both the vibration and airborne models fail to accurately predict the depth of the IL trough at resonance. The airborne model indicates that at this frequency the important material parameters controlling the response are the foam loss factor η_f and the air flow resistivity R_1. An explanation for the discrepancy between modelling and measurement is sought in these two parameters.

8.2.1 Air Flow Resistivity

The air flow resistivity of a porous material, R_1 (Nsm^{-4} or rayl/m) is defined by [26])

$$R_1 = (dp/dx)/v$$

where dp/dx is the pressure differential across a sample of thickness dx and v the steady, or *dc*, average linear flow velocity through unit cross section area perpendicular to the flow direction. The value of R_1 is obtained from the pressure gradient dp/dx for $v \rightarrow 0$ and assumes that the flow into and out of the porous material is unrestricted. The parameter R_1, which will now be designated as R_1 (*dc*), is incorporated into the airborne IL model via Equation (29).

For *dc* flow through PU foams the pressure gradient dp/dx increases with increasing flow velocity [39], and although it is often assumed that R_1 (*dc*) is applicable to the interaction between air and the frame this is not entirely correct [36] and in some cases the flow resistivity at acoustic frequencies can be considerably larger than the *dc* value [37]. The effective air flow resistivity R_1 (*ac*) can be derived from R_1 (*dc*) using the expression [38]

$$R_1\,(ac) = R_1\,(dc)F_r(\varkappa)$$

where $F_r\,(\varkappa)$ is the real part of a mathematically complex function $F(\varkappa)$. The parameter $\varkappa = a_p\,(\omega/\nu)$ where ν is the viscosity of air and a_p has dimensions of length and is related to the mean pore radius of the foam. Estimating a value for a_p we can calculate the frequency correction factor for the flow resistivity, i.e. relate R_1 (*ac*) to the measured value R_1 (*dc*), using the charts given by Biot [39]. For a pore radius of 0.25 mm the correction factors are given approximately as

Frequency	$\varkappa$	Correction Factor
100	1.5	1.0
1000	4.6	1.4

A correction factor for R_1 (*ac*) of 1.2 is given by Bies [26] which is intermediate between the two values calculated here.

The Biot model used to calculate the value of $\varkappa$ consists of an array of nonconnecting tubes. When interconnectivity is taken into account he showed that the correction factor is $\tau\varkappa$ where he called τ the sinuosity. We take this as being equivalent to the tortuosity of the foam frame. Taking the simplest case where the average particle flow vector is at 45° to the mean flow direction the value of τ = 1.4 [30] which gives asrongection factor of about 1.6 at 1 kHz. The Biot approach accounts only for Poiseuille flow but flow through flexible foams can be represented by the non-linear Forscheimer fluid flow equation which contains two terms [31]; the friction term and a flow inertia term, which is strongly dependent on flow velocity. Thus we might expect that the correction factor at acoustic frequencies would be larger than that given by the Biot equation, i.e. > 1.6.

Measured values of R_1 for the foam cores used in our tests ranged from ~50–100 kNsm^{-4}. In the model predictions, Figures 13 and 14, we have used a value of R_1 *(ac)* = 200 kNsm^{-4}. It is seen that even if the correction factor for R_1 is set > 2 the predicted depth of the IL resonance trough is larger than that measured. It is concluded that the difference between modelling and measurement cannot be totally accounted for in terms of the effective flow resistivity.

*8.2.2 **Panel Damping***

The model developed in Section 3.2, based on the Beranek and Work acoustic transmission line concept, assumes that the sound waves are incident normally on the metal panel, the impervious barriers, i.e. panel and septum, are of infinite extent, the motion of the barriers is not constrained in any way, so they move in a piston like fashion, and that they behave as simple limp masses. None of these conditions obtain for the system configuration used in the test rig and they do not apply to practical situations. For example the test panel is finite, clamped around its boundaries (4C), exhibits a high modal density and as discussed in Section 3.1, the motion of the panel is damped. Panel damping results from the inherent loss of the metal and radiation damping, which are relatively small, the clamped boundaries which are sealed using a lossy elastomer and the interaction of the panel with the foam core. Vibration theory indicates that system damping primarily affects behaviour under resonance conditions so it is not unreasonable to presume that the poor agreement between modelling and measurement at system resonance is associated with the damping of the metal panel by the carpet composite.

Au and Byrne [25] have addressed this problem. Starting with the

Beranek and Work model they derived an expression, equivalent to Equation (20), for the acoustic impedance of a damped panel on the incident side. This is

$$Z_i = Z_1 + j(\omega m_1 + K^*k^4/\omega)$$

where $K^* = K'(1 + j\eta)$ is the complex stiffness of the panel and k $(=2\pi/\lambda)$ is the wave number of the bending waves in the panel. The incorporation of this expression into the transmission line model might account for differences between measurement and modelling at system resonance.

REFERENCES

1. Eeckhaut, G., T. Pham and R. J. Lockwood. 1987. "Improved Sound Insulation in Motor Vehicles," *Polyurethanes World Congress,* Sept. 29–Oct. 2, pp. 708–711.
2. Berthevas, P. R., A. Fanget and G. Gatouillat. 1987. "The Development of a Sound Insulation Package for Car Floor Coverings Using a Combination of Polyurethane Technologies," *Polyurethanes World Congress,* Sept. 29–Oct. 2, pp. 701–707.
3. Zwinselman, J. J. and W. D. Bachmann. 1988. "Polyurethane Foams for Sound and Vibration Damping in Automotive Applications," *J. Cellular Plast.,* 24:274–283.
4. Gahlau, H. K. 1987. "Advanced Noise Control in Automotive Engineering," *Polyurethanes World Congress,* Sept. 29–Oct. 2, pp. 712–726.
5. Lee, J. R. 1988. "Automotive Sound Insulation–The Role of Polyurethanes," *UTECH88,* The Hague, Oct. 11–13, pp. 81–82.
6. Private communication.
7. Hilyard, N. C., D. McFarland, A. M. Elliot and A. Cunningham. 1990. "Tailor Made all MDI Based Moulded Floor Systems for the Automotive Industry," *33rd Annual Polyurethane Technical/Marketing Conference,* Sept. 30–Oct. 3, pp. 631–638.
8. Hilyard, N. C. and A. Cunningham. 1991. "Design Optimisation of MDI Based Polyurethane Foam Backed Automotive Carpet Systems," *Proc. SAE Noise and Vibration Conference,* Traverse City, pp. 435–441.
9. Ling, M. K., N. C. Hilyard and A. Cunningham. 1991. "Fundamental Studies of the Acoustic Behaviour of Foam Backed Floor Covering," *Proc. Cellular Polymers International Conference,* London: Rapra Technology Ltd., March 20–22.
10. Ling, M. K. 1993. "Vehicle Interior Noise Control Using Three Layer Composites with Polyurethane Cores," *J. Sound Vib.,* 106:103–109.
11. Cunningham, A. and N. Duggan. 1994. "Advances in Polyurethane for Use in Acoustic Applications for the Automotive Sector," UTECH94, The Hague, March 22–24, paper 7.

12. Rhodes, D. M., A. V. Phillips and B. Abbot. 1988. "Prediction of Engine Vibration from Combustion Pressure Measurements and Examination of the Effect of Combustion Variations in the Time Domain," *Proc. I. Mech. E.*, C33/88, pp. 147–157.
13. Jha, S. K. 1976. "Characteristics and Sources of Noise and Vibration and Their Control in Motor Cars," *J. Sound Vib.*, 47(4):543–557.
14. Nefski, D. J. and Sung, S. H. 1985. "Vehicle Interior Acoustic Design Using Finite Element Methods," *Int. J. Vehicle Design*, 6(1):24–40.
15. V'er, I. L. and C. I. Holmer. 1971. *Noise and Vibration Control*, L. L. Beranek, ed., McGraw-Hill, Chap. 11.
16. Cremer, L., M. Heckle and E. E. Ungar. 1988. *Structure Borne Sound, 2nd Edition*, Springer Verlag, (a) p. 461, (b) p. 482, (c) p. 323, (d) p. 498 and (e) p. 459.
17. Beranek, L. L. and G. A. Work. 1949. "Sound Transmission Through Multiple Structures Containing Flexible Blankets," *J. Acoust. Soc. Am.*, 21(4):419–428.
18. Beranek, L. L. 1947. "Acoustical Properties of Homogeneous, Isotropic Rigid Tiles and Flexible Blankets," *J. Acoust. Soc. Am.*, 19(4):556–568.
19. Lauriks, W. 1994. "Acoustic Characteristics of Low Density Foams," *Low Density Cellular Plastics—Physical Basis of Behaviour*, N. C. Hilyard and A. Cunningham, eds., Chapman and Hall, Chap. 10.
20. Bolton, J. S. 1984. Ph.D. Thesis, *Cepstral Techniques in the Measurement of Acoustic Reflection Coefficients with Application to the Determination of Acoustic Properties of Elastic Porous Materials*, University of Southampton.
21. Plunt, J. 1991. "The Power Injection Method for Vibration Damping Determination of Body Panels with Applied Damping Treatments and Trim," *Proc. SAE Noise and Vibration Conference*, Traverse City, pp. 417–425.
22. Ling, M. K., N. C. Hilyard and R. B. W. Heng. 1990. "Performance of Polyurethane Cored Carpet Composites for Vehicle Interior Noise Control," *Proc. Inst. Aoustics 90 Conference*, Southampton, March 27–30, 12(1):307–312.
23. Mangiarotty, R. A. 1963. "Optimization of the Mass Distribution and Air Spaces in Multiple-Element Soundproofing Structures," *J. Acoust. Soc. Am.*, 35(7):1023–1029.
24. Byrne, K. P. 1984. "Prediction of the Performance of Acoustic Lagging," *Internoise 84*, Honolulu, Dec. 3–5, pp. 425–428.
25. Au, A. C. K. and K. P. Byrne. 1987. "On the Insertion Loss Produced by Plane Acoustic Lagging Structures," *J. Acoust. Soc. Am.*, 82(4):1325–1333.
26. Bies, D. A. 1971. "Acoustical Properties of Porous Materials," *Noise and Vibration Control*, L. L. Beranek, ed., McGraw-Hill, Chapter 10.
27. Gibson, L. J. and M. F. Ashby. 1988. *Cellular Solids: Properties and Structure*, Pergamon Press, Chapter 5.
28. Gent, A. N. and A. G. Thomas. 1963. "Mechanics of Foamed Elastic Materials," *Rubb. Chem. Tech.*, 36:597–610.
29. Attenborough, K. 1981. "Acoustical Characteristics of Porous Materials," *Physics Reports*, North Holland Publishing Company, Section 2.2, 82(3):179–227.

30. Hilyard, N. C. and P. Collier. 1987. "A Structural Model for Air Flow in PUR Flexible Foams," *Cellular Polymers,* 6(6):9–26.
31. Morris-Kirby, R. 1991. "The Optimisation of Acoustic Polyurethanes Using the Computer Aided Noise Analysis and Simulation Programme (C.A.N.A.S.)," *Polyurethanes World Congress,* Sept. 24–26, pp. 257–265.
32. Klingenberg, H. 1988. *Automobil-Messtechnik,* BandA, Acoustik, Springer-Verlag.
33. Interkeller, AG, Zurich, 1974.
34. Nichols, R. H., H. P. Sleeper, R. L. Wallace and H. L. Ericson. 1947. "Acoustical Materials and Acoustical Treatment for Aircraft," *J. Acoust. Soc. Am.,* 19(3):428–443.
35. BS2750-1993 (ISO140) Measurement of Sound Insulation in Building Elements.
36. Ingaard, K. U. 1981. "Locally and Non-Locally Reacting Flexible Porous Layers: A Comparison of Acoustical Properties," *Trans. ASME,* 103:302–313.
37. Schroeder, F. K. 1953. "Der Schallschluckgrad als Funktion des Schalleinfallswinkels," *Acoustica,* 3:54–66.
38. Lambert, R. F. 1982. "The Acoustic Structure of Open Cell Foams," *J. Acoust. Soc. Am.,* 72(3):879–887.
39. Biot, M. A. 1955. "Theory of Propagation of Elastic Waves in Fluid Saturated Porous Solids: II. Higher Frequency Range," *J. Acoust. Soc. Am.,* 28(2):179–191.

Pentane Blown Polyurethane Foams

O. Volkert
BASF AG
67056 Ludwigshafen, Germany

ABSTRACT

HCFCs (hydrochlorofluorocarbons) were developed as a CFC (chlorofluorocarbon) replacement but considered only as a short-term solution since their ODP (ozone depletion potential) is not zero.

With HFCs (hydrofluorocarbons), which are considered as the next generation of blowing agents, a basic problem arises: in order to be non-flammable and to have a low thermal conductivity the fluorine content of the molecule must be high. This means, however, high production costs and an increased GWP (greenhouse warming potential).

For this reason a consequent change to halogen-free hydrocarbons has taken place in Europe. Pentane isomers were chosen as the only suitable candidates due to their boiling points.

In flexible integral skin foam *n*-pentane was introduced already 4–5 years ago. Density distribution and mechanical foam properties of R11 and *n*-pentane blown integral skin foams are nearly identical.

Rigid foams, where thermal conductivity and low permeation of the cell gas from the foam are important properties, must be considered more individually. In sandwich panels for the construction industry a higher thermal conductivity can, if necessary, be compensated for by increasing the thickness of the part. In this area less expensive *n*-pentane and isopentane are used.

In the appliance field an increase in wall thickness is hardly possible. Here cyclopentane is used, whose thermal conductivity is similar to that of the best HFCs.

Permeation of all pentane isomers from PUR-foam is as low as that of R11, resulting in a similar ageing behavior. Also similar are thermal conductivity at low temperature and mechanical properties of R11 and pentane blown foams.

The increased flammability of pentane foams can be overcome by adding additional flame retardant and by proper adaption of the foam formulation. Safe procedures have been developed and are offered for the processing of flammable blowing agents.

In Europe, cyclopentane has been used in refrigerators for the past 2 years and *n*-pentane has been used in flexible integral skin foams for 4–5 years without any major problems.

1. INTRODUCTION

ALTHOUGH IT HAS BEEN PREVIOUSLY SHOWN THAT PENTANE IS A SUITable blowing agent for polyurethane foams [1], it never really made the breakthrough compared to expanded polystyrene (EPS) where pentane has always been used. The main reason for this was undoubtedly its flammability, but apparently ideal alternatives were also available at an early stage in the form of CFCs. This situation changed with the development of the ozone debate. Since CFCs had to be phased out, intensive efforts went into the search for substitutes. At great expense, the so-called HCFCs were developed. However, it soon became clear that these were only temporary alternatives. Because the reaction of HCFCs with OH free radicals in the troposphere is faster than that of CFCs, their concentration in the stratosphere and thus their ozone depletion potential (ODP) is considerably lower but not zero. Public opinion has now become so sensitive on this issue that only an ODP of 0 would be acceptable in a long term solution. The next generation of blowing agents currently under development is the hydrofluorocarbons (HFCs). However, they have a fundamental problem: fluorine-rich HFCs, like hexafluorobutane [2] or dodecafluorohexane [3] for example, have a high molecular weight and thus relatively low thermal conductivity and are also non-flammable. However, the more fluorine atoms in the molecule, the more expensive the product and the higher its greenhouse warming potential (GWP), since its reaction rate with OH free radicals is lower and the number of IR-absorbing C–F groups increases. On the other hand, if the number of fluorine atoms is reduced, as in difluoroethane (R 152a) or trifluoroethane (R 143a) [4], the thermal conductivity increases, the products become more volatile (usually gaseous at room temperature) and become flammable. Additionally, there is the toxicity problem of partly fluorinated hydrocarbons, which can hardly be predicted for a distinct compound thus causing long and extensive toxicity studies. It was an obvious step to move to hydrocarbons, which are completely free of fluorine.

It is interesting that this substitution has been carried out rigorously in Europe, while the response in North America has been more cautious. Virtually all development work and examples of the use of pentanes reported below therefore originate in Europe.

2. PENTANES AS BLOWING AGENTS

Owing to the boiling point, the only hydrocarbons actually suitable as blowing agents are pentanes. Cyclopentane and *n*-pentane/isopentane differ so greatly in their property profile that—irrespective of the requirements in the foam—one or the other of the isomers is employed specifically (a further isomer, neopentane, is not available in industrial quantities and is therefore negligible).

The flashpoint of all pentane isomers is below −20°C, i.e. they are flammable (Table 1).

Availability

Isopentane and *n*-pentane have long been employed for the production of expanded polystyrene and are therefore available in industrial quantities. A mixture of 80 parts by weight of *n*-pentane and 20 parts by weight of isopentane is frequently used, the combination known as pentane S. Cyclopentane is available in 2 different purity grades: about 95%, i.e. virtually pure cyclopentane, and 75%, which contains an additional 10% of *n*-pentane and 15% of 2,2-dimethylbutane. Both grades are used by processors.

3. PENTANE IN INTEGRAL SKIN FOAM

In integral skin foam, the temperature gradient which results in the foam due to the good thermal conduction at the metallic mold is utilized to produce a corresponding density profile [5]. The crucial factor here is the boiling point of the blowing agent under the pressure prevailing in the mold. At the surface of the foam part, the boiling point is above the prevailing temperature. A compact, unfoamed skin is thus obtained, but it is below the foaming temperature inside the part, resulting in a foam core of lower density.

If the matrix is elastic, flexible integral skin foams are obtained, which are used to produce shoe soles or automotive parts such as arm-

Table 1. Physical properties of pentane isomers.

		n-Pentane	Isopentane	Cyclopentane	R11
Boiling point	(°C)	36	28	49.5	24
Vapor pressure at 20°C	(bar)	0.57	0.78	0.35	0.89
Density at 20°C	(g/ml)	0.626	0.620	0.745	1.488
Thermal conductivity at 25°C	(mW/mK)	14.8	14.8	12.1	8.0
Explosion limits in air	(% by vol.)	1.4–7.8	1.3–7.6	1.4–8	—

rests, headrests and steering wheels. Thermosets are produced if the matrix is highly crosslinked and rigid. Typical applications of thermosets are window frames and technical casing parts.

It has been found that *n*-pentane is the ideal replacement for R11 in flexible integral skin foams. In thermosets, the heat balance is different. Higher-blowing agents are required due to higher reaction temperatures occurring in the mold.

Pentane in Flexible Integral Skin Foams

The proportion of R11 in the polyol component is 5–7% by weight for flexible integral skin foams. *n*-Pentane has about half the molecular weight of R11 and, since the gas yields of the two blowing agents are virtually the same, only about 3% by weight of *n*-pentane is required in the polyol to give the same foam densities.

Foam Properties

Using *n*-pentane as blowing agent, both skin formation in the foam and the mechanical properties of the parts are comparable to those made using R11.

Figure 1 shows the density distribution of the corresponding shoe soles. Table 2 shows the mechanical values of test sheets for a polyether shoe sole system.

Since the processing behaviour of both systems is nearly identical, the only drawback to the R11 replacement is the flammability of pentane.

Safety Problems in the Finished Part

Since flexible structural foams are open-cell foams, the majority (about 90%) of the pentane escapes during foaming in the mold and during demolding. Of the remaining 10%, about 3/4 diffuses out of the part within 10 hours. Only a small amount remains included in the compact skin, from where it diffuses out of the foam part very slowly over the course of weeks or months. Figure 2 shows the pentane concentrations, determined by gas chromatography, in a shoe sole during storage. The theoretical pentane content in the sole has been calculated to be 1.14%.

The amount of pentane measured 10 minutes after demolding was only 0.21% in the sole. After 8–10 hours, the pentane content had

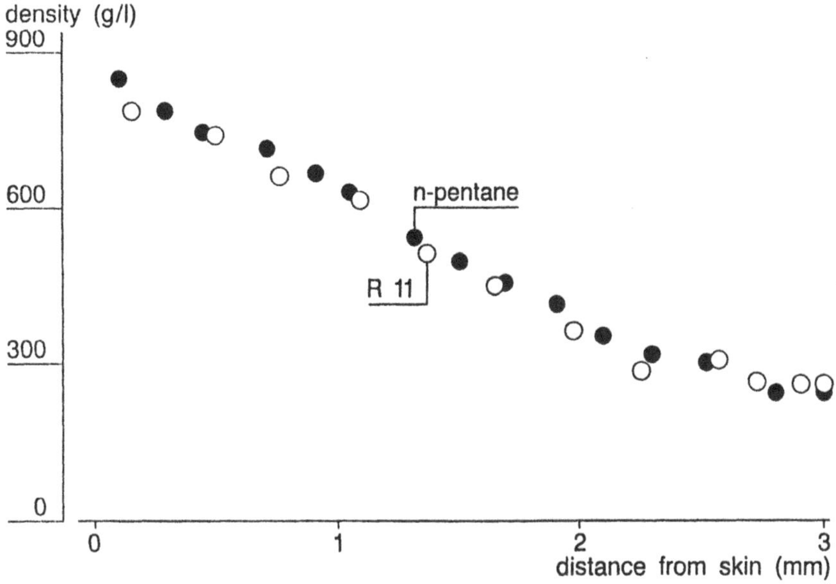

Figure 1. Density distribution of an integral skin foam.

dropped to 450 ppm, due to the pentane, which had been encapsulated in the structural skin.

After 8–10 hours in a suitably ventilated room, sufficient amounts of pentane had escaped so that the residual content is negligible. Explosive pentane/air mixtures are thereby avoided so that the parts can be processed further without safety hazards.

In larger parts, for instance in automotive applications the diffusion of pentane from the parts takes longer, and molded parts should therefore be stored in facilities with proper ventilation.

Table 2. Mechanical properties of a shoe-sole-system.

Blowing Agent (% in polyol)	2.88 *n*-Pentane	5.50 R11
Free rise density (g/l)	190.0	200
Molded density (g/l)	550	550
Hardness (Shore A)	64	65
Tensile strength (N/mm^2)	5.4	4.9
Tear strength (N/mm^2)	8.3	8.3
Elongation (%)	410	400

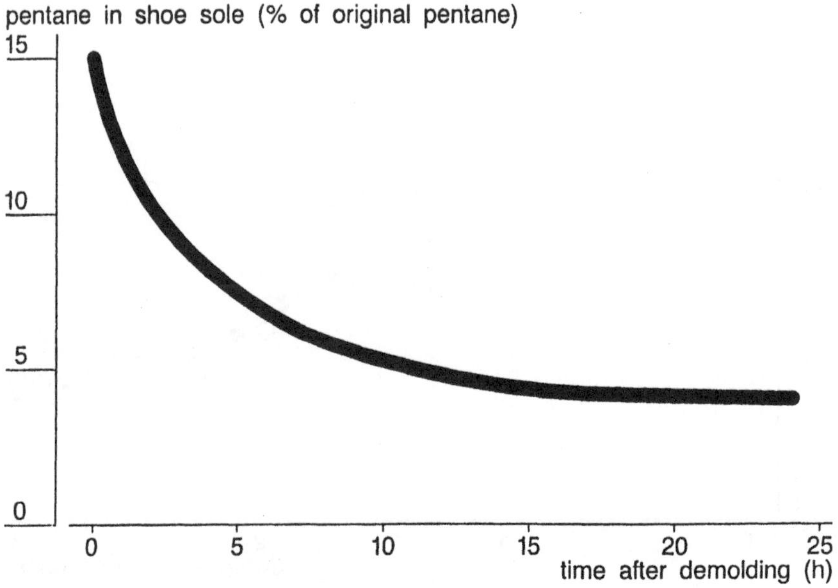

Figure 2. Diffusion of n-pentane from an integral skin foam.

4. PENTANE IN RIGID FOAM

Rigid polyurethane foams are generally used as insulating materials. A low thermal conductivity of the blowing gas is required since it provides insulation to the foam. Since outer layers which prevent diffusion are not present in many applications but insulating effects are to be retained for many years, the blowing gas must remain in the cells during this period, i.e. its diffusion rate through the polyurethane matrix must be low. Although the gas phase accounts for only about 50 to 60% of the thermal conductivity of the foam (the remainder is due to the matrix and radiation), a higher thermal conductivity of the cell gas has a serious effect on the insulation properties of the material. Higher conductivity can, if necessary, be compensated in sandwich elements by increasing the thickness of the element. This, however, is not possible in refrigerators. An increase in the wall thickness would directly reduce the useful volume since the outer dimensions of a refrigerator are usually standardized and therefore cannot be changed. In addition, an increase in the foam thickness would require new molds which are costly.

The use of pentane in rigid foam can therefore be divided into two separate areas: cyclopentane as the gas of lowest thermal conductivity for refrigerators and *n*-pentane/isopentane for applications in the construction sector.

Cyclopentane in Refrigerators

Blowing Efficiency

In spite of its rather high boiling point, cyclopentane has a high gas yield in rigid foams. About 26 l/mol were measured in a typical appliance formulation (Figure 3). The maximum value obtained theoretically is 30–31 l/mol (the reaction temperature at the gel point is about 100 °C).

Thermal Conductivity

The aim was to achieve the thermal conductivity of foams containing 50% less R11, the foam system used until 1993. The thermal conductivity of these foams was 19 mW/mK at 25 °C. Cyclopentane has a 50% higher thermal conductivity than R11 (Table 1). The difference in thermal conductivity in the foam is less since the proportion of CO_2 in the

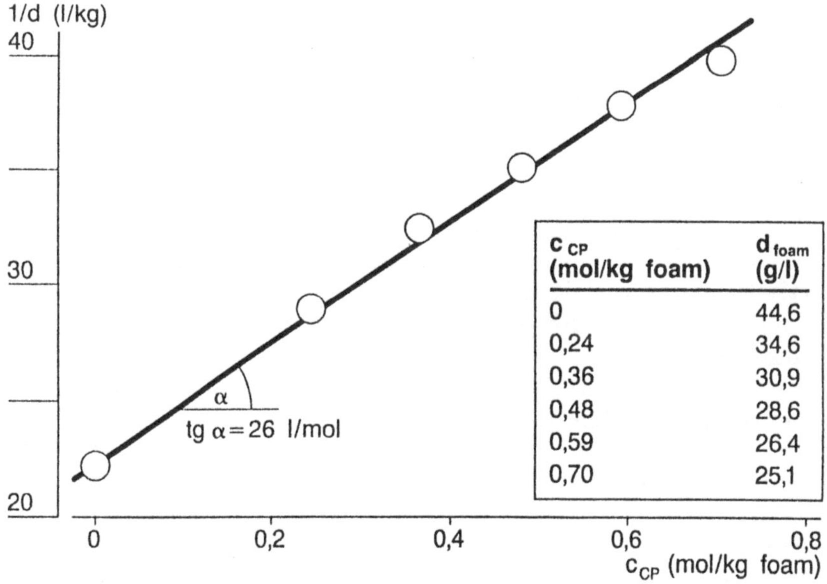

c_{CP} (mol/kg foam)	d_{foam} (g/l)
0	44,6
0,24	34,6
0,36	30,9
0,48	28,6
0,59	26,4
0,70	25,1

Figure 3. Gas yield of cyclopentane blown rigid foam.

R11 systems is somewhat greater and, as mentioned above, the gas phase only contributes about 60% of the total conductivity of the foam. An accurate calculation gives an increased conductivity of cyclopentane-blown foam of 6–8%, which is also found in direct comparative experiments. By adjusting the polyurethane formulations to suit the new blowing agent and by optimizing the cyclopentane:CO_2 ratio in the cell gas, the 19 mW/mK target has now been attained and even improved.

Ageing of Foams

The thermal conductivity of rigid polyurethane foam changes in time due to diffusion and dissolution processes: CO_2 diffuses out of the foam relatively quickly and air diffuses into the foam. Since the thermal conductivity of air is greater than that of CO_2, the conductivity of the system as a whole increases due to these diffusion processes and reaches a constant final value when diffusion is complete. The partial pressure of CO_2 is then zero, and that of air is 1 bar. Figure 4 shows the course of such diffusion processes with time in a rigid polyurethane foam sample.

In elements 2–3 cm in thickness, the half life of CO_2 is about 10 days and that of air is about 12 weeks.

Besides these diffusion processes, heavy gas (cyclopentane or R11) dissolves in the matrix, a process that occurs approximately as long as air diffusion. The result is that heavy gas disappears from the gas phase, and the heavy gas:air ratio drops, thus increasing the thermal conductivity (to a first approximation, the thermal conductivity depends on the composition of the cell gas and not on the pressure in the cell).

Comparative ageing values of the thermal conductivity of R11- and cyclopentane-blown foams show that the curves intersect. Although the R11 variant has a somewhat lower initial conductivity the aged conductivity of the cyclopentane foam is somewhat lower than that of the R11 foam when diffusion and dissolution processes are complete (Figure 5).

The drop in the partial pressure of R11 due to dissolution processes is evidently greater than that of cyclopentane.

Long-Term Ageing

Even after CO_2/air diffusion is complete, the thermal conductivity continues to increase slowly, since the heavy gases also diffuse out of the foam. This causes the heavy gas:air ratio in the cells to drop fur-

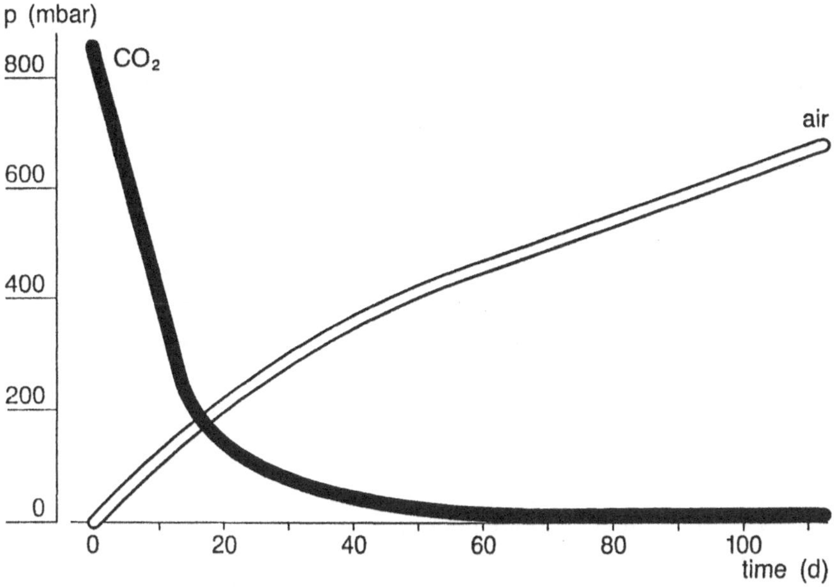

Figure 4. *Diffusion of CO_2 and air in rigid foam.*

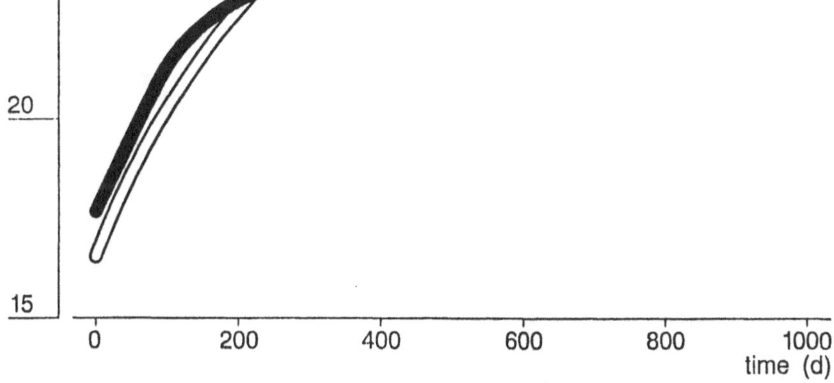

Figure 5. *Ageing of cyclopentane and R11 blown rigid foam.*

ther, resulting in an increase in thermal conductivity. However, the diffusion rate of cyclopentane is several orders of magnitude lower than that of CO_2 or air, so that at room temperature these processes can only be detected by measurements taken over many years. For this reason, accelerated ageing tests have been carried out at elevated temperature (80°C), showing that the diffusion coefficient of cyclopentane is approximately the same as that of R11 [6]. This means that cyclopentane has a half life of 30–50 years in parts which are 2–3 cm thick. Since the average service life of refrigerators in Europe is about 10 years, cyclopentane diffusion has virtually no practical significance.

Low-Temperature Behavior

The boiling point of cyclopentane is 49.5°C, which is at the upper limit for a practical blowing agent. The vapor pressure at room temperature is only about 350 mbar and at lower temperatures it drops further (Figure 6).

At −20°C, the usual temperature in freezers, it is only 50 mbar. The consequent drop in the cyclopentane:CO_2 or cyclopentane:air ratio should theoretically result in a significant increase in the thermal conductivity of the foam.

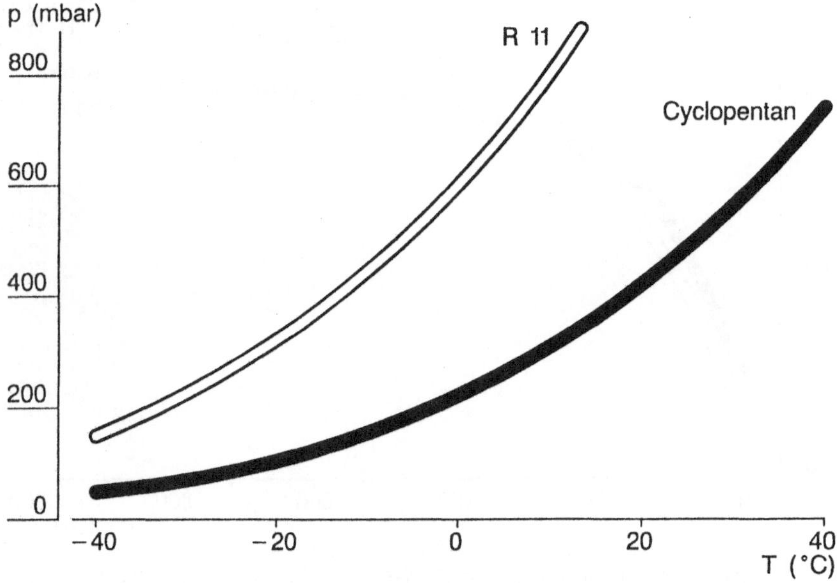

Figure 6. Vapour pressure of cyclopentane and R11.

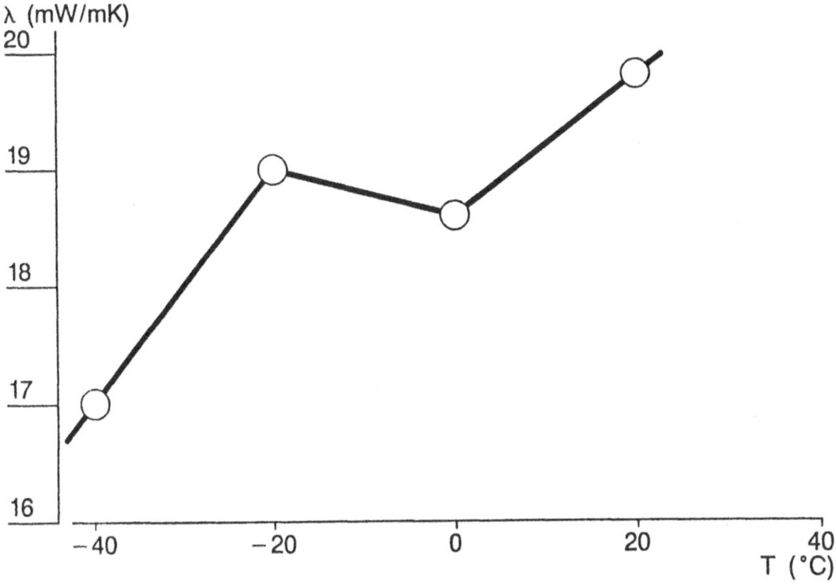

Figure 7. Temperature dependence of thermal conductivity of CP-blown foam.

However, experimental measurements have indicated a decrease also at lower temperatures (Figure 7).

Apparently, the dominant factor is the temperature dependence of the other criteria in thermal conductivity (matrix, radiation, but also the cell gas itself), so that the change in cell gas has no adverse effect.

Mechanical Properties of the Foams

At room temperature, 50% of the cyclopentane used (about 4.4% by weight based on the foam) condenses in the cells since the equilibrium vapor pressure is only 350 mbar. The Henry solubility coefficient for cyclopentane in a rigid polyurethane foam matrix has been found to be about 1 mol/kg bar [7], so that the condensed amount corresponds approximately to the amount dissolved at 350 mbar and the cell internal pressure does not drop any further due to dissolution processes.

However, this low vapor pressure of the cyclopentane means that the total internal pressure in the cells is about 200 mbar lower than that for comparable R11 formulations. For this reason, cyclopentane foams are more sensitive to shrinkage, and the molding densities employed are about 10% higher than in the case of R11. The mechanical properties and the shrinkage are then satisfactory.

Table 3 shows typical values obtained during final expansion of a brett mold (200 × 20 × 5 cm).

Flammability of the Foam

There are no official fire regulations for refrigerator foams. In order to be sure that the cyclopentane-blown foams in refrigerators are not a fire hazard, flammability tests were carried out on cyclopentane- and R11-blown foams. The pure foams and elements made from polystyrene-laminated foam have been tested.

In the conventional foam test (DIN 4102), cyclopentane and R11 foams gave virtually the same result when the entire element was tested. In the case of unlaminated foam, the flammability of the cyclopentane foam was slightly higher. Evidently, the small amount (4%) of blowing agent in the material affects the flammability much less than expected.

Pentane-Blown Foams for the Construction Industry

Strictly speaking, a distinction should be made between construction elements in the form of insulating boards (density range 30–35 g/l, flexible outer layers of aluminum, GRP or cardboard) and sandwich panels (density 40–45 g/l, steel outer layers). However, since the basic problems using pentane are very similar, no further differentiation is made here.

At present, the majority of the polyurethane foam employed in the construction industry is blown using *n*-pentane and isopentane. The 80:20 mixture of the two isomers (pentane S) is frequently also used. Since the properties of the two isomers are very similar (see Table 1), no distinction will be made between the two compounds when referring to pentane.

Table 3. Foam properties of a cyclopentane blown rigid foam for refrigerators.

Density (10% overpacking)	(g/l)	38.9
Closed cells	(%)	94
Compression strength	(kPa)	170
Dimensional stability	(Vol-%)	
−30°C		±0.1
+80°C		±0.5

Pentane Versus Cyclopentane

In Germany, polyurethane insulating foam for the construction industry is divided into certain insulation classes known as thermal conductivity groups. Assignment to a particular group is made on the basis of the thermal conductivity of the foam plus defined additions which represent ageing processes of the foam from experience. A thermal conductivity group covers a range of 5 mW/mK. Although cyclopentane foams have a lower thermal conductivity than pentane foams, this is not sufficient for classification in the next-lowest thermal conductivity group.

Since foams blown with the two blowing agents are classified in the same insulation class, the price is therefore the determining factor in favor of pentane. The current price ratio between pentane and cyclopentane is about 1:3.

A further advantage of pentane over cyclopentane is its lower boiling point and higher vapor pressure. A higher vapor pressure of the blowing agent means a higher pressure in the cells and thus a lower tendency toward shrinkage. This also undoubtedly has a positive effect on the compression strength.

Solubility of Pentane in Polyols

A major disadvantage of pentane is its relatively low solubility in conventional rigid foam polyols. In order to obtain the low foam densities required, the water content in the polyol must be increased. This results in a higher viscosity of the polyol component and thus lower miscibility with the isocyanate and poorer flow properties of the system. In order to counteract this, polyols of lower viscosity have been developed, usually at the expense of functionality. This in turn causes lower crosslinking density and thus lower compression strength and dimensional stability.

It can thus be seen that pentane is in no way a straightforward drop-in substitute for R11 in this area of application, but that the foam formulations must be modified correspondingly on the polyol side.

Thermal Conductivity

In sandwich panels produced using pentane as a blowing agent in an optimized system, the thermal conductivities achieved are between 21 and 22 mW/mK.

As already outlined as refrigerator problems, the thermal conductivity in outer layers open to diffusion increases due to diffusion of CO_2 and air.

The increase in the case of pentane foams is substantially parallel to that of R11 foams, i.e. the difference of about 4 mW/mK in the initial conductivity remains constant during ageing. The final value obtained after completion of CO_2/air diffusion is about 25 mW/mK.

In outer layers which are closed to diffusion, the thermal conductivity remains constant (22 mW/mK).

Long-Term Ageing

In order to estimate long-term ageing, i.e. the diffusion of pentane out of the foam, effective diffusion coefficients of pentane and R11 were measured at elevated temperature (80 °C) [8]. The results are shown in Table 4.

The results show that diffusion proceeds at an equally slow rate for all three compounds. Assuming that the 80 °C values can be extrapolated back to room temperature for all 3 compounds, the half life of pentane is about 50 years in parts having a thickness of 5 cm, so that pentane diffusion is negligible as in the case of cyclopentane in refrigerators.

Mechanical Properties

Pentane has lower solubility than cyclopentane not only in the polyols but also in the polyurethane matrix of the resultant foam. However, in contrast to the polyol solubility, this poor matrix solubility has advantages with respect to the mechanical properties of the foam. Readily soluble blowing agents act as plasticizers in the polyurethane matrix as well as lower the internal pressure in the cells. Both effects impair the compression strength of the foam.

On the other hand, a disadvantage of the compressive strength of pentane foams is the frequent consequence for a higher concentration of flame-retarding agents, which likewise act as plasticizers, compared with R11 foam. In order to compensate for this, the density of pentane

Table 4. Effective diffusion coefficients (80 °C) in rigid PUR-foam.

R11	5.5 10^{-8} cm^2s
n-Pentane	9.1 10^{-8} cm^2s
Isopentane	3.5 10^{-8} cm^2s

Table 5. Properties of pentane and R11 blown rigid foams for sandwich elements.

Blowing Agent		R11	Pentane
Amount of blowing agent (parts/100 parts of polyol)		14	7
Density	(g/l)	38	41
Compression strength	(kPa)	135	130
Dimensional stability	(%)		
−30°C		<1	<1
+80°C		<1	<1
Thermal conductivity (initial value)	(mW/mK at RT)	17	21–22

foams is therefore usually set 10% higher than that of R11 foams. A comparison of typical thermal and mechanical properties of pentane- and R11-blown foams for sandwich elements is shown in Table 5.

Flammability of the Foam

In contrast to the situation in refrigerators, flammability regulations do exist for polyurethane foams in the construction industry. However, the regulations frequently differ from country to country, and the problem is therefore very difficult to deal with on the whole. In general, it can be stated that although pentane, in contrast to R11, is a highly flammable substance, the differences in flammability of the corresponding foams are surprisingly small. Polyisocyanurate foams give nearly the same results with pentane and R11 if the same amount of flame retardant is used. In polyurethane foams, the amount of flame retardant must be increased if pentane is used. Additionally, further measures are often taken to keep the increase in the amount of flame retardant within limits: the NCO:OH ratio is increased, i.e. small amounts of isocyanurate structures are produced, and at least some of the polyether-polyol is replaced by polyester-polyol. Pentane foams then exhibit similar behavior to R11 foams in many tests (i.e. small burner test according to DIN 4102, mini-corner test, oxygen index, cone calorimeter test, fire shaft test, epiradiateur test, etc.). Detailed results are given in the patents and references in the literature [9–15].

5. SAFETY

Pentanes are flammable. Their most important properties with respect to safety are shown in Table 6.

Table 6. Safety data of pentane.

		n-Pentane	Isopentane	Cyclopentane
Flash point	(°C)	< −20	< −20	< −20
Ignition point	(°C)	285	420	380
Explosion limits in air	(% by vol.)	1.4–7.8	1.3–7.6	1.4–8.0

As stated earlier, pentane-containing foams represent virtually no safety risk: structural foams contain no pentane after only a relatively short time and in the case of rigid foams, the pentane is well encapsulated in the individual foam cells. The time taken for significant amounts to diffuse into the atmosphere is so long that pentane cannot be detected in the ambient air.

The situation is different during the production of foams, which involves handling of liquid pentane.

In order to reduce risks to a minimum, pentane is usually not added until the polyol component is being processed. Only in the case of flexible structural foams are there some small processors who purchase the finished, pentane-containing polyol component.

The problems which occur during processing are similar for all pentane isomers, and therefore no distinction will be made here between different substances or types of application.

The risks can be divided into the following areas:

1. Storage of pentane
2. Preparation of pentane-containing polyol components
3. Problems in the foaming machine
4. Problems during foaming

Storage of Pentane

The supply, storage and conveying of pentane are subject to safety standards, as prescribed, for example, in gas stations at a reasonable expense.

Preparation of the Pentane-Containing Polyol Component

Pentane is incorporated into the polyol in a premix tank. The pumps used are either leakproof or monitored for leaks. If possible, the tank should be accommodated in a separate, well-ventilated housing.

Problems in the Foaming Machine

Fixed pipelines which have been tested for leaks are used for conveying the ready-to-use polyol component. Shut-off valves which react automatically to any drop in pressure minimize risks in the pipelines. The interior of the foam machine silo, into which the pentane-containing solution is pumped, is flushed with nitrogen. In order to prevent pressure losses due to the conventional stirrer seals, the tank contents can also be mixed by means of a magnetic stirrer or a circulating pump.

In critical areas, gas sensors monitor the air quality in order to detect leaks. If specified limits are reached, warning signals are activated or plant parts are switched without current. There is therefore no need to equip large parts of the plant with expensive explosion-proof electrical fittings.

It is simpler to circumvent the problem of premix tank and pentane-containing polyol in the silo by metering pentane directly into the polyol component via a static mixer just before the mixing head of the foaming machine. Although this method is used frequently in continuous plants, it causes considerable problems in batch processes. Only the premix tank method is therefore used in the refrigerator industry, for example.

Problems During Foaming

On exiting the mixing head of the foaming machine and during subsequent removal, pentane is emitted and must be extracted, for example in the case of twin-belt plants. The situation is more critical in the production of refrigerators, where prefabricated plastic shapes are expanded in a closed metal mold. The loss of cyclopentane has been measured in model experiments as being 2–5% of the cyclopentane employed. The cyclopentane concentration right at the flowing foam front is higher than the upper explosion limit, and only a short distance away it already drops below the lower explosion limit, i.e. both positions are non-critical. In between, however, the concentration passes through the critical area, and if electric discharges between charged plastic parts should occur at this point, an explosion or at least a deflagration could theoretically occur (probably the latter, since the amount of ignitable mixture is low). Absolute safety is therefore only guaranteed if either the mold is flushed with nitrogen or electric discharge is prevented by appropriate antistatic finishing of the plastic material (some machinery suppliers already offer mixing heads with an additional line for nitrogen). In practice, major refrigerator manufacturers in Europe have worked for almost 1-1/2 years without these safety precautions and without a single accident. Therefore the situation in practice seems to be significantly less critical than expected theoretically.

In the meantime, suitable concepts for the production of pentane-containing polyurethane foams are now also available from all manufacturers of polyurethane machines [16–18].

6. ENVIRONMENTAL POLLUTION

The ozone depletion potential of pentanes is zero, since they do not contain chlorine. They react very quickly with OH free radicals in the troposphere and are degraded. Table 7 shows measured rate constants for the reaction of *n*-pentane and cyclopentane with OH free radicals [19] and the resulting half lives calculated for the two compounds in the troposphere.

These short half lives mean that the greenhouse warming potential corresponds, to a first approximation, to the CO_2 value of its carbon content, i.e. it is virtually zero.

Pentanes are volatile organic compounds (VOCs) and under certain circumstances (high NO_x, dust and moisture contents in the air and high light energy) can contribute to smog. With respect to the VOC problem, a distinction should be made between two factors:

1. The photochemical ozone creation potential (POCP) as a property of a VOC
2. The amount of VOC emitted

Regarding the POCP, C_3–C_5-alkanes are classified as VOCs of relatively minor importance. Hydrocarbons having >6 carbon atoms are more relevant [20].

The amount of emitted pentane is low (2–5% of the total amount measured in refrigerators). The regulations concerning pentane use in Europe are easily met. In the United States the situation is more com-

Table 7. Half-life of pentane in the troposphere.

	k(OH + pentane)	$t_{1/2}$ (pentane)
n-Pentane	4.31 10^{-12} cm^3s^{-1}	3.72 d
Cyclopentane	5.35 10^{-12} cm^3s^{-1}	2.98 d

Table 8. Toxicity of pentane.

		n-Pentane	Isopentane	Cyclopentane
MAC	(ppm)	1000	1000	—
TLV	(ppm)	600	—	600
LD_{50}, oral	(mg/kg)	>16,000	>16,000	>2000
LD_{50}, dermal	(mg/kg)	>2500	>2500	>2500
LD_{50}, inhalation	(mg/l)	>100	>100	>25
Mutageneity		no	no	no
Carcinogeneity		no	no	no

plicated. For any new plant the specific regulations have to be fulfilled and discussed with the corresponding authorities. Pentanes are biodegradable in water and soil, where their half lives are estimated at < 100 days.

7. TOXICOLOGY

Table 8 shows an overview of the data currently available on the toxicity of pentane.

Isopentane and *n*-pentane are compounds which have already been used for a long time in expanded polystyrene. Their toxicity has therefore been investigated thoroughly and no problems have been encountered.

Some information is still lacking regarding cyclopentane. Cyclopentane is a naturally occurring compound (petroleum) and has been used in gasoline for years. It does not react with PUR foams and is not decomposed by constituents of these foams. Migration studies on cyclopentane-containing foams showed negligible concentrations of cyclopentane in refrigerators. As far as we know today, cyclopentane is thus not a particular toxicological risk, and there is certainly no need for a comprehensive toxicity study. Nevertheless, to be on the safe side some supplementary toxicological studies are currently being carried out. A summary of this work will be published shortly on its completion.

8. CONCLUSIONS

Pentane isomers are suitable replacements for CFCs and HCFCs in polyurethane foams. Because the ODP and GWP of these blowing agents are zero they are not considered as an interim but as an optimum solution.

Using pentane as a blowing agent, the thermal, mechanical and processing properties of the resulting foams are good. The slightly higher flammability of the foams can be overcome by proper formulations. The safety hazards encountered during foam production have been minimized.

In flexible integral skin foam, *n*-pentane has been in use for 4–5 years. In rigid foams either cyclopentane, *n*-pentane or isopentane are now being used, depending on the foam requirements. The European appliance industry has meanwhile changed to using cyclopentane as a blowing agent. In the Far East the change is underway. In the construction field introduction of pentane use proceeds more slowly due to time consuming official testing requirements. It is assumed, however, that the change will soon occur also in this area.

REFERENCES

1. 1965. *Encyclopedia of Polymer Science and Technology*, 2:532–534.
2. 1991. E-PS 344 537, Bayer AG, *PUR World Congress*, W. M. Lamberts, Nice.
3. E-PS 507 214, Höchst AG.
4. Barthelemy, P. P., A. Leroy, L. Zipfel, W. Krücke and J. A. Fanklin. 1993. *PUR World Congress*, Vancouver.
5. Piechota, H. and H. Röhr. 1975. *Integralschaumstoffe*, Carl-Hanser-Verlag München.
6. Volkert, O. 1993. *PUR World Congress*, Vancouver.
7. BASF AG, unpublished results.
8. BASF AG, unpublished results.
9. Heilig, G. and R. E. Wiedermann. 1993. *PUR World Congress*, Vancouver.
10. Heilig, G. 1991. *Kunststoffe*, 81:622–625.
11. Heilig, G., F. H. Prager, R. Walter, R. Wiedermann and F.-W. Wittbecker. 1992. *Bauphysik*, 14:76–81.
12. Heilig, G., F. H. Prager, R. Walter, R. Wiedermann and F.-W. Wittbecker. 1991. *Kunststoffe*, 81:790–794.
13. Demharter, A. 1994. *PUR-Hartschaum im Bauwesen*, SKZ Würzburg.
14. Nicola, W. J. and H. U. Weber. 1994. *PUR World Congress*, Boston.
15. Riley, R. E., T. L. Fishback and L. C. Yu-Hallada. 1994. *PUR World Congress*, Boston.
16. 1993. Data Sheet, Cannon-Group, Milano.
17. Mariani, V. and P. J. Allgaier. 1994. *PUR World Congress*, Boston.
18. Taverna, M. and P. Corradi. 1994. *PUR World Congress*, Boston.
19. Güsten, H. and L. Klassinc. 1986. *Naturwissenschaften*, 73:129–135.
20. 1992. Report TNO Inst. of Environmental and Energy Technology.

Applications of Polyols Derived from Renewable Resources in Polyurethanes

SHAIO-WEN WONG AND SONG-BO WANG
Polymer Institute
University of Detroit Mercy
Detroit, MI 48219-0900, U.S.A.

JAMES F. GRAVES AND INNOCENT ODOCHA
Department of Biology
Center of Excellence in Environmental Engineering and Science
University of Detroit Mercy
Detroit, MI 48219-0900, U.S.A.

ABSTRACT

Lesquerella oil and castor oil were investigated in the synthesis of polyurethanes and copolymers of urethane-urea and urethane-isocyanurate. These polymer systems can be utilized in the preparation of elastomers, sealants, adhesives, and composites.

In the preparation of polyurethanes, methylene-4,4′-di(phenyl isocyanate) (MDI), either the solid monomer or the liquid modified MDI, and 2,4-/2,6-toluene diisocyanate (TDI) are often used to link the polyols and hydroxyl chain extenders. To simulate the performance of castor oil and lesquerella oil in actual production, a model reaction was designed using a mixture of phenyl and *o*-tolyl isocyanate to react with the vegetable oils. Besides urethane formation, urea, trimer and other compounds can also be produced in the reaction. The selectivity of these reactions was determined by means of high performance liquid chromatography (HPLC).

Polyurethane and polyurethane-isocyanurate elastomers based on castor and lesquerella oils were prepared using a one shot process. The mechanical and thermal properties and morphologies of these elastomers were measured. The biodegradability of these polymers was examined using the bacterial strain *Pseudomonas aeruginosa* ATCC 13388 and five strains of fungi.

INTRODUCTION

THE ULTIMATE SOURCE OF ALL LIFE AND ALL NATURAL PRODUCTS IS THE sun. Mankind has depended on the constantly renewable biomass as its primary source for material needs since prehistoric times. We are concerned here with primarily the conversion of biomass to marketable industrial products. The known processes used to derive chemicals from natural products are inefficient and expensive because of the need to isolate the desired products from solutions of relatively low concentrations. With the arrival of relatively low-cost and high-volume petrochemicals, most of the industries based on natural products were replaced. Very few polymers have their origins in natural products and limited research and development have continued only in the areas in which raw materials are readily available, such as the by-products and wastes from the forestry industry. For example, sebacic acid is commercially derived from castor oil; it forms the "10" component of nylon 6,10. Products from castor oil in Third World countries such as India, Brazil, and China, in which castor oil is produced in abundance locally, can compete with the high-tech petrochemical industry.

All vegetable oils are triglycerides of fatty acids. Although most of these triglycerides contain only double bond functional groups, a few oils naturally contain other groups such as hydroxyl in castor oil [1], safflower oil [2] and lesquerella oil [3] or oxirane in vernonia oil [3]. The castor plant is grown commercially worldwide, especially in India, Brazil, and China. Lesquerella grows in the desert regions of the western United States, particularly in Arizona. Vernonia grows in a broad expanse throughout central Africa. All three of these oils contain functional groups, hydroxyl or epoxy, which make them rare among the huge number of triglyceride oils. The hydroxyl groups in these oils can react with di- or polyfunctional isocyanates to form branched polyurethanes and the epoxy groups can react with isocyanates to form oxazolidones. Sperling et al. [4,5] reported the preparation of simultaneous interpenetrating networks based on vernonia oil polyesters and polystyrene. New polymers can be synthesized from these oils, and a new industry could be brought to the specific area that produces these plants. This would also relieve some pressure on the dependence of petrochemicals in our industry.

Currently, only the castor oil derivatives are commercially available. Due to their unique structure, castor oil derivatives provide better water resistance and are also lower in viscosity than the polyether and polyester polyols. They exhibit exceptional ability in pigment-wetting and penetration due to the presence of the ricinoleate. The castor oil–based polyurethane systems have been widely used in the telecommunication/electrical industries, coatings, adhesives and sealants applications [6]. They have also served the biomedical field in such critical areas as blood and human tissue contact [1]. However, relatively few studies have been carried out on the structure-property relationships of castor oil–based polyurethanes [7–12]. A fast curing polyurethane-urea system [13] based on castor oil, isophorone diisocyanate (IPDI) and methylene-4,4′-dianiline (MDA) has the potential to be processed by reaction injection molding (RIM). Only a few studies were reported on the preparation of polyurethanes using other vegetable oil derivatives such as hydroxymethylated linseed oil [14] and safflower oil [15].

A resin composition [16] containing isocyanurate rings was prepared by heating hexamethylene diisocyanate and hydrogenated castor oil with a trimerization catalyst, *N,N,N,*-trimethyl-*N*-2-hydroxylammonium *p*-tert-butylbenzoate. A coating made from this resin with a petroleum hydrocarbon solvent has low toxicity, is easy to work with and gives good weather resistance.

A series of UV curable castor oil–based polyurethane networks [17] was prepared by treating castor oil with TDI and hydroxyethyl methacrylate in various ratios in the presence of various reactive diluents and photosensitizers. The mechanical properties of the products depended on the reactant molar ratio, the content and nature of the reactive diluents and the hard segment content. These materials exhibited excellent hydrolytic stability.

There is increasing concern about waste plastics in the environment. The plastic products comprise a growing share of the general litter and the associated solid waste disposal problems as they take over a greater share of the markets. Mounting pressure from the public has forced federal and state legislature to consider mandating plastic degradability.

The microbial degradation of polyurethanes has been investigated by several groups [18–21]. Darby and Kaplan [18] concluded that polyester-based polyurethanes are much more susceptible to degradation than polyurethanes derived from polyether diols. It was discovered in a study [21] of microbial degradation of caprolactone-based polyurethanes that two fungi, *Aspergillus fumigatus* and *Fusarium solanii* and a yeast, *Cryptococcus Laurentii* were effective in degrading the polyurethanes. In general, the more flexible the polyurethanes, the more susceptible they are to biodegradation. The regularity of the arrangement of the soft and hard segments in the molecular structure and the rigidity of the segments also play a very important role in the degradation of the material. Polyurethanes consisting of rigid, regularly arranged segments are more likely to be resistant to biodegradation. It is anticipated that the polymers derived from vegetation (polyesters of fatty acids) are more susceptible to degradation by hydrolysis, heat, oxidation, photodegradation and biodegradation.

In the preparation of polyurethanes, methylene 4,4′-di(phenyl isocyanate) (MDI), either the solid monomer or the liquid modified MDI, and 2,4-/2,6-toluene diisocyanate (TDI) are often used to link the polyols and hydroxyl chain extenders. To simulate the performance of castor oil and lesquerella oil in actual production, a model reaction was designed using a mixture of phenyl and *o*-tolyl isocyanate to react with the vegetable oils. The reactivity of DB castor oil and lesquerella oil with isocyanate in the presence of various catalysts was compared with that of *n*-butanol. The selectivity of various isocyanate reactions was determined by means of high performance liquid chromatography (HPLC). Due to the complexity of lesquerella oil and DB oil composition, only the extent of the OH−NCO reaction and trimer formation was determined.

Polyurethane and polyurethane-isocyanurate elastomers based on castor oil, lesquerella oil, a carbodiimide-modified MDI and 1,4-butanediol were prepared using a one-shot process. The mechanical properties and morphology of these elastomers were determined. Biodegradability of these polymers was tested by the growth of bacteria (ASTM G22 agar method using bacterial strain *Pseudomonas aeruginosa* ATCC 13388) and fungi (ASTM G21 method with five ATCC strains of fungi).

EXPERIMENTAL

Reactivity Studies of Model Reactions

The model reactions were carried out in a 300-ml three-neck round bottom flask, equipped with a magnetic stirring device and kept under a constant flow of nitrogen. The reaction flask was placed in a constant-temperature oil bath, and 50 ml of a diglyme solution mixture of phenyl isocyanate/*o*-tolyl isocyanate were pipetted into the flask. The isocyanate content of the solution was determined by DBA titration. The titration method was a modified version of ASTM D-1638. The toluene-DBA solution and aqueous HCl were used in the titration. The hydroxyl component and catalyst were carefully weighed into a 50 ml volumetric flask that was then diluted to the 50 ml mark with diglyme. This flask was also immersed in the oil bath. When the temperature of the two components reached 50°C, the solution containing the hydroxyl component was poured into the reaction flask. Zero time of the reaction was taken when half of the solution was emptied. At various time intervals, two liquid samples were taken from the reaction solution. One sample was quenched with a DBA/toluene solution that was used to determine the NCO content by back titration with standardized HCl solution. The second sample was added to the *p*-dioxane-DBA solution that was analyzed by HPLC to determine the concentration of various species existing in the solution.

HPLC Analysis

Synthesis of Model Compounds

The model compounds of the anticipated reaction products were synthesized by reacting phenyl isocyanate and *o*-tolyl isocyanate with the respective compounds using various catalysts. The reactions were carried out in diglyme solution at temperatures ranging from 40° to 80°C. Diglyme (bp 162°C) was removed from the mixture by vacuum distillation at 40 mmHg and 60°C after the isocyanate was completely consumed. These purified compounds were characterized using element analysis, FTIR, NMR, and melting point determination. The purity of these compounds was checked by HPLC analysis. The synthesis of these model compounds is summarized in Table 1.

HPLC Analysis

A Perkin Elmer LC-65T HPLC was used for the analysis. Separation of these species in dioxane solution containing di-*n*-butylamine was achieved by employing a NOVA-PAK CN radial-PAK cartridge (8 mm × 10 cm, Waters) and a UV spectrophotometer as the detector (254 nm). A 50 μl sample loop was connected to a Dynamax autosampler (Rainin Model AI-2) to ensure the same volume injection for each sample. Two solutions were used to analyze these compounds since the extinction coefficient of trimer and carbamates based on DB oil and lesquerella oil is much lower than that of the other compounds. The eluent consists of a mixture of *p*-dioxane and *n*-heptane (15/85 volume ratio) at a flow rate of 1 ml/min. The concentrations of the samples for the analysis of trimer and carbamates based on DB and lesquerella oils were about 5–10 times higher than those for the other compounds.

Data Analysis

A Macintosh Classic II computer connected to a Dynamax MacIntegrator II (Rainin) was used for the data acquisition and analysis. The

Table 1. Synthesis of model compounds.

Designation	OH/NH_2, eq	Reactants NCO, eq	Cat. wt%	Reaction Temp., °C	Yield, %	mp, °C
BPC	*n*-BuOH, 0.07	PhNCO, 0.07	—	50	95	60
BTC	*n*-BuOH, 0.07	*o*-TNCO, 0.07	Dabco T, 1.1	50	95	40
DBPC	DB oil, 0.03	PhNCO, 0.03	Dabco T, 0.8	45–50	91	brown, reddish liq.
DBTC	DB oil, 0.03	*o*-TNCO, 0.03	Dabco T, 0.9	45–50	79	brown, reddish liq.
LPC	Les. oil, 0.03	PhNCO, 0.03	Dabco T, 0.9	45–50	67	brownish, red liq.
LTC	Les. oil, 0.03	*o*-TNCO, 0.03	Dabco T, 0.8	45–50	78	brownish, red liq.
DBAPU	DBA, 0.05	PhNCO, 0.05	—	25	100	83–84
DBATU	DBA, 0.05	*o*-TNCO, 0.05	—	25	50	pale brownish liq.
WPU	H_2O, 0.45	PhNCO, 0.1	—	25	100	241–242
WTU	H_2O, 0.45	*o*-TNCO, 0.1	Dabco 33LV, 0.2	25	62	256–267
TPI	—	PhNCO, 0.06	Dabco TMR-3, 1.0	80	60	278–279
TTI	—	*o*-TNCO, 0.06	Dabco TMR-3, 1.0	80	5	253 white
					54	148 brown

Notes: PhNCO = phenyl isocyanate; *o*-TNCO = *o*-tolyl isocyanate; *n*-BuOH = *n*-butanol; Les. oil = lesquerella oil; and DBA = di(*n*-butyl)amine.

retention time, peak height and peak area for each species were recorded.

Preparation of Elastomers by the One-Shot Method

The calculated amounts of polyol, hydroxyl chain extender and catalyst were mixed thoroughly in a plastic cup. The liquid carbodiimide modified MDI (Isonate 143L, Dow) was then added to the mixture with vigorous stirring for 30 seconds. The mixture was poured into a preheated mold (50–55 °C). At the moment the mixture started to gel, the mold was closed and a pressure of 10,000 psi was applied to the mold. Samples were demolded after one hour and were post-cured for 20 hours at 90–100 °C.

Measurement of Mechanical Properties

The stress-strain properties of the elastomers were measured after the test specimens were conditioned at room temperature and 50% humidity for at least one week. The procedure was based on ASTM D412-80. Uniaxial stress-strain properties were measured at room temperature. The properties measured include 100% secant modulus, 300% secant modulus, ultimate tensile strength, and elongation at break. A microdie was used to cut specimens from the test sheets. Testing was performed on an Instron Tester (Table Model A-1020). The sample thickness ranged from 0.070 to 0.090 inches. The crosshead speed was 5 in. per min. The initial distance between the grips (0.6 inch) was used to calculate the elongation at break. The reported data are the average of at least 3 readings.

Thermal Analysis

TA-9900 thermoanalyzer (TMA and DSC) from TA Instruments was used to measure the thermal transition temperatures such as glass transition, softening temperature and melting point. The thermal stability of the elastomers was measured using a TGA.

Thermomechanical Analysis (TMA)

The TMA measurement was carried out at a heating rate of 10 °C/min. under a tray weight of 5–50 gram using the penetration mode. Samples were heated from −150 °C to the softening temperature.

Differential Scanning Calorimeter (DSC)

The DSC was programmed to heat from −150° to 280°C at a rate of 10°C/min. under nitrogen atmosphere (a flow rate of 60 cc/min.). The sample weight ranged from 5 to 8 milligrams.

Thermogravimetric Analysis (TGA)

The TGA experiment was carried out under nitrogen atmosphere at a flow rate of 40 ml/min. The heating rate was 10°C/min., and the sample size ranged from 8 to 15 mg.

Biodegradability Testing

ASTM G22, which uses the bacterial strain *Pseudomonas aeruginosa* ATCC 13388, has been utilized in this study. This test appeared to be good for screening large numbers of polymer samples for biodegradability by this one microorganism. Polymer samples based on castor oil and lesquerella oil underwent incubation with bacteria.

ASTM G21, which uses five strains of fungi, was also utilized to test the biodegradability of the polymer. The five fungi, *Aspergillus niger* (ATCC 9642), *Penicillium funiculosum* (ATCC 9644), *Chaetomium globosum* (ATCC 6205), *Gliocladium virens* (ATCC 9645), *Aureobasidium pullulans* (ATCC 9348), were received from the American Type Culture Collection (ATCC). They were rehydrated from a freeze-dried state and cultured on Sabouraud Dextrose Agar. A quality control characteristic that has become apparent is that they can be distinguished from each other by their pigmentation. Polymer samples were screened by this biodegradability test method. Microbial growth curves (yields) on polymer samples in liquid nutrient salts medium in flasks on a water bath shaker produced the best quantitative data on the growth of microorganisms on the polymers. The polymer was the only carbon source available to microorganisms in this assay.

RESULTS AND DISCUSSION

Reactivity of DB Oil and Lesquerella Oil with Isocyanate in the Presence of Various Catalysts and Other Components

In the preparation of polyurethanes, methylene 4,4′-di(phenyl isocyanate) (MDI), either the solid monomer or the liquid modified MDI, and 2,4-/2,6-toluene diisocyanate (TDI) are often used to link the polyols and hydroxyl chain extenders. To simulate the performance of castor oil and lesquerella oil in actual production, a model reaction was designed using a mixture of phenyl and *o*-tolyl isocyanate to react with the vegetable oils. Diglyme was chosen as the solvent since the alkyl ether group is the dominant component in a polyether-based polyurethane. The speed of the reaction can be determined from the disappearance of isocyanate in the solution. The reactivity of DB castor oil and lesquerella oil with isocyanate in the presence of various catalysts and other components was compared with that of *n*-butanol. The reactions were carried out at 50°C. The isocyanate consisted of a mixture of phenyl and *o*-tolyl isocyanate in a volume ratio of 8:2. The rate of reaction was followed by the disappearance of NCO groups. The reactions described below were studied.

Urethane Formation in the Presence of Catalyst Dabco T

Figure 1 shows the NCO disappearance as a function of time in the reaction of isocyanates with hydroxyl-containing compounds in diglyme at 50°C catalyzed with tertiary amine Dabco T. The same equivalents of NCO and OH components were used in the reaction and the concentration of catalyst was 1% based on the weight of the OH

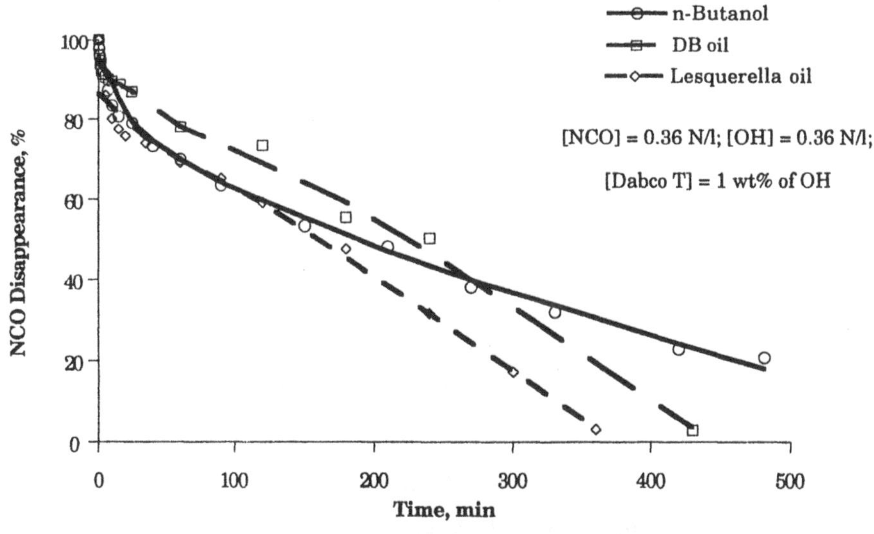

Figure 1. Isocyanate reactions of hydroxyl compounds and phenyl/o-tolyl isocyanate (80/20) in diglyme at 50°C catalyzed with tertiary amine Dabco T.

component. The results indicate that the reactivity of these OH-compounds could be made comparable using different amounts of catalyst. Bear in mind that the equivalent weights of DB oil and lesquerella oil are much higher than that of *n*-butanol. The amount of catalyst used in the reactions of DB oil and lesquerella oil was much higher than that used in the reaction of *n*-butanol. In these catalyzed reactions, the initial rate of lesquerella oil was comparable to that of *n*-butanol, but became slightly faster as the reaction proceeded. The disappearance of NCO in the reactions of lesquerella oil and castor oil was faster than that of *n*-butanol after 50% conversion. The impurities in vegetable oils might cause the faster NCO consumption, since only the moisture in vegetable oils had been removed under vacuum (80°C and 5 mmHg overnight).

Urethane-Isocyanurate Formation in the Presence of Catalysts Dabco T and Dabco TMR-3

The rate of the three OH-compounds was about the same in the urethane-trimerization reactions as shown in Figure 2. Excess isocyanate (equivalent ratio of NCO/OH was 2/1) was used in these reactions in order to form isocyanurate under the influence of a trimeriza-

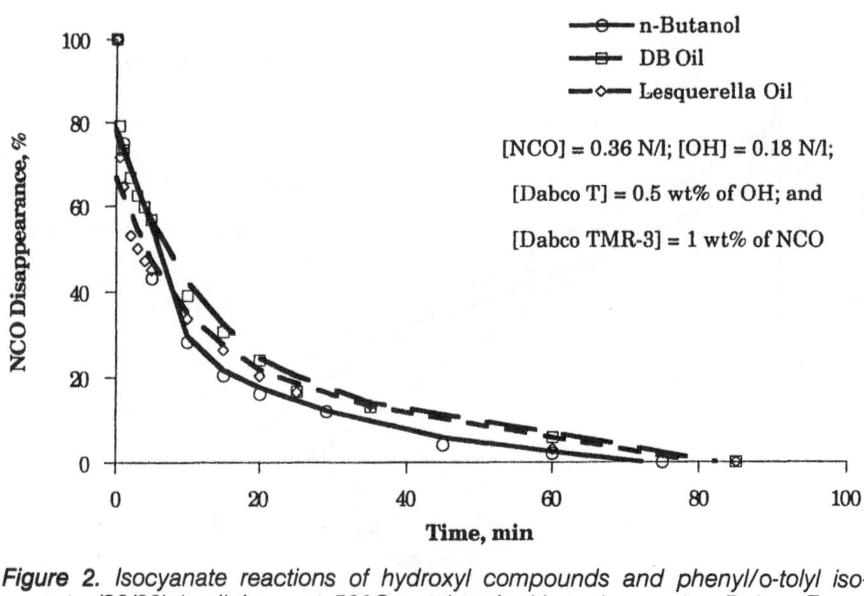

Figure 2. Isocyanate reactions of hydroxyl compounds and phenyl/o-tolyl isocyanate (80/20) in diglyme at 50°C catalyzed with tertiary amine Dabco T and quaternary ammonium carboxylate Dabco TMR-3.

tion catalyst Dabco TMR-3. The concentrations of Dabco T and Dabco TMR-3 were 0.5% based on the weight of the OH component and 1% based on the weight of NCO.

Selectivity of Carbamate and Trimer Formation in Competing Isocyanate Reactions

High performance liquid chromatography (HPLC) was used to analyze the concentration of various species in the reaction solutions that were quenched in di-*n*-butylamine dioxane solution at different times. The selectivity of carbamate, urea, and trimer formations in these reactions could then be determined. Six reactions were analyzed. They were urethane formation in the presence of Dabco T and urethane-isocyanurate formation in the presence of Dabco T and Dabco TMR-3.

Synthesis of Model Compounds

The reactivity study from the NCO disappearance results in only the overall consumption of isocyanates. Isocyanate is a very active group that can react with many compounds containing active hydrogen such as alcohols, amines, and water. It can also react with itself to form dimer or trimer. The selectivity of these reactions was determined by HPLC. To perform the HPLC analysis, compounds anticipated in the reactions had to be synthesized.

The following model compounds were thus synthesized:

- urethanes: The products of phenyl isocyanate with DB castor oil, lesquerella oil and *n*-butanol; and *o*-tolyl isocyanate with DB castor oil, lesquerella oil and *n*-butanol
- ureas: The products of phenyl isocyanate with water and di-*n*-butylamine; and *o*-tolyl isocyanate with water and di-*n*-butylamine
- isocyanurates: phenyl isocyanate and *o*-tolyl isocyanate trimerization products

The syntheses were carried out in diglyme solution using different catalysts to accelerate the respective reactions. The structure of the samples was characterized using Fourier Transform Infrared Spectroscopy (FTIR). The melting point of the solid samples was determined by Differential Scanning Calorimeter (DSC) (Table 1).

Concentration Calibration

To determine the retention time and peak height (area) of the model compounds, several DBA-dioxane solutions containing different concentrations of the species were used to establish the calibrations of the concentration of each species with their respective peak heights or peak areas. However, due to the complexity of the chromatograms of DB oil/NCO and lesquerella oil/NCO, the peak height was used for the calculation. Figures 3–5 show the chromatograms of the compounds anticipated in the reactions of *n*-butanol, DB oil, and lesquerella oil, respectively. The calibration curves between the concentration and peak

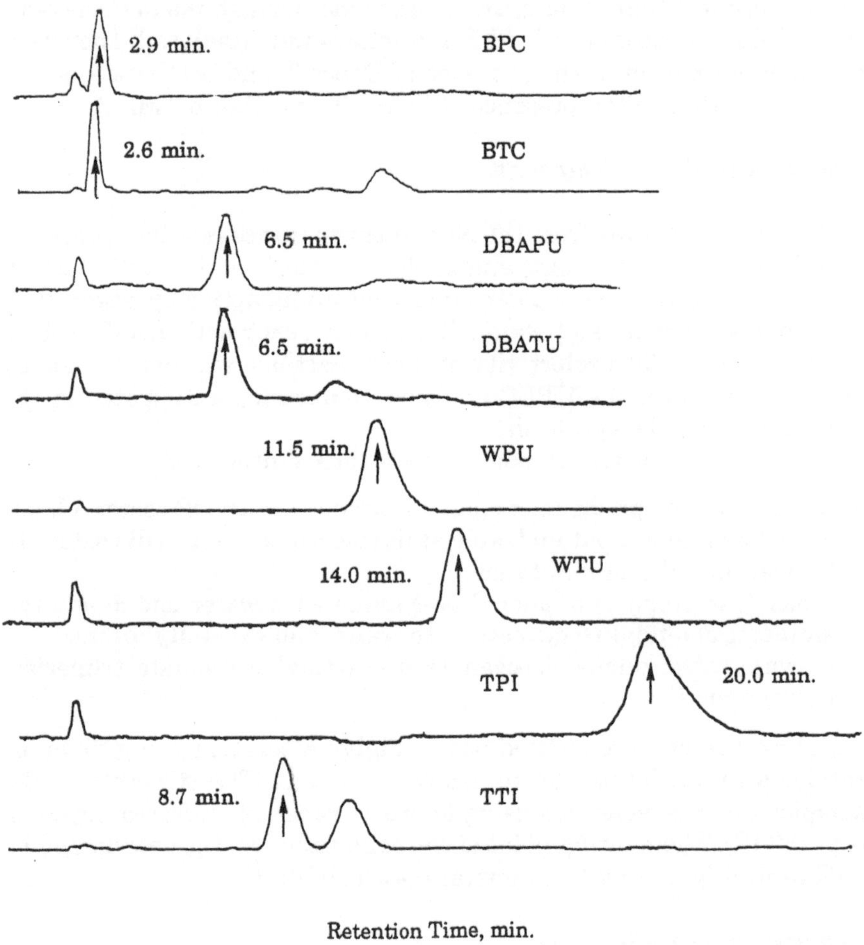

Figure 3. Chromatograms of possible compounds in the reaction of n-butanol *and phenyl/o-tolyl isocyanate.*

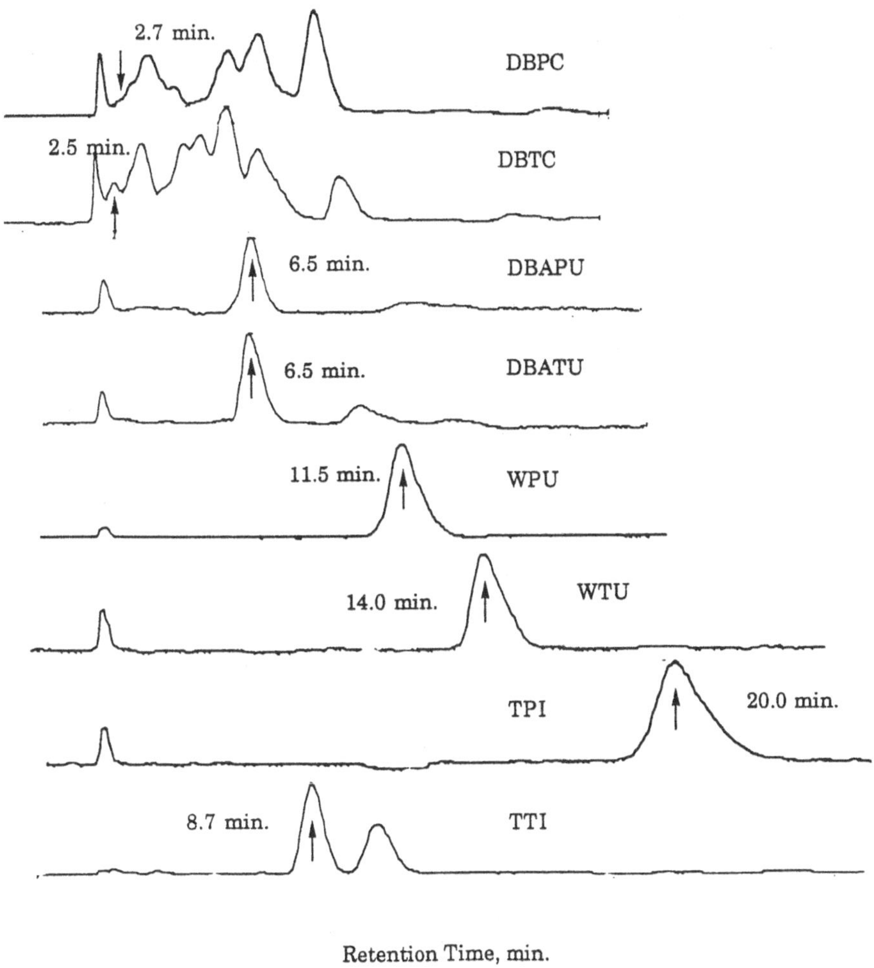

Figure 4. Chromatograms of possible compounds in the reaction of DB oil and phenyl/o-tolyl isocyanate.

height at a specific retention time for each compound was then established. For some compounds such as the products of DB oil and lesquerella oil with isocyanates, their HPLC chromatograms were very complicated. It seemed that the peak height of all the peaks was linearly correlated with the concentration of the species. However, in the actual analysis, most of the peaks were overlapping with the peaks of other compounds appearing in the solution. The peaks chosen for the calculation were those in which no interference occurred in the chromatograms of the reaction samples. The concentration of various species in the reaction solution could be calculated from the equations listed in Table 2.

Concentration Calculation

It was noted that all the chromatograms exhibited a peak at a retention time of 2 minutes (H_2). The height of this peak remained more or less constant. We used this peak as an internal reference for the calculations. Although an autosampler was used to inject 50 μl samples for the analysis, the peak height of the H_2 peak varied from sample to sam-

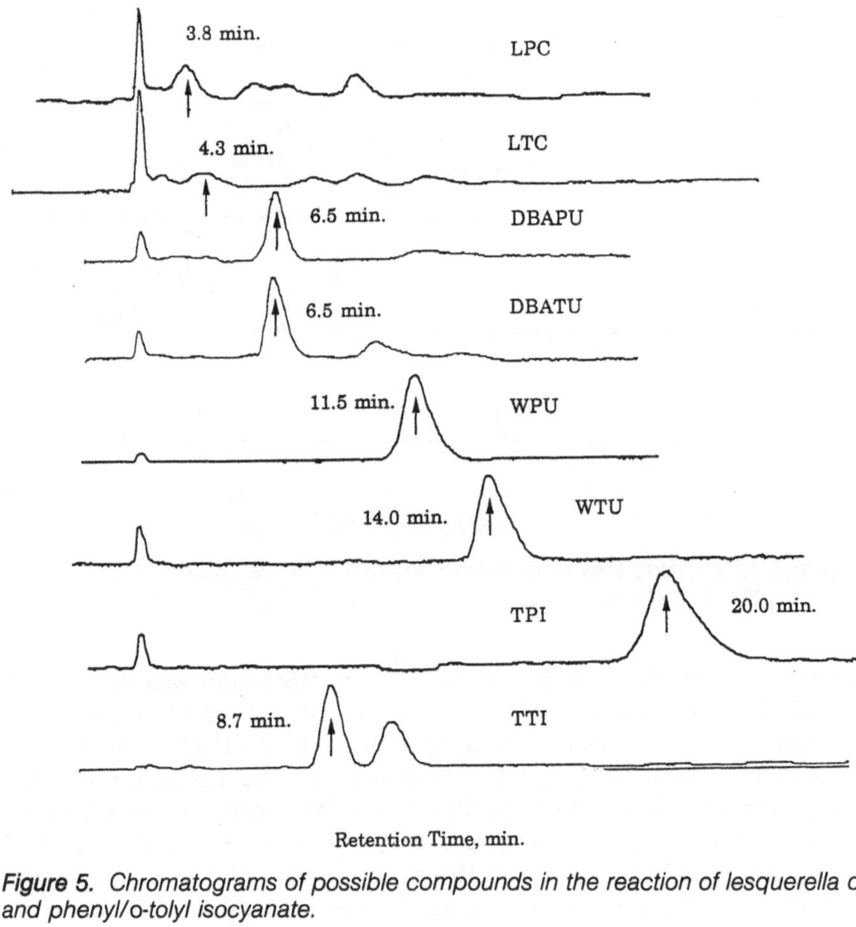

Figure 5. Chromatograms of possible compounds in the reaction of lesquerella oil and phenyl/o-tolyl isocyanate.

Table 2. Peak height versus concentration of model compounds.

Compound	Retention Time, min.	Linear Regression Equation, Peak Height vs. Conc.	Conc. Range, g/g × 10^5	Correlation Coefficient
BPC	2.9	$H_{2.9} = -298 + 2674[BPC] \times 10^5$	12–475	0.999
BTC	2.6	$H_{2.6} = 13417 + 1432[BTC] \times 10^5$	10–500	0.968
BDPC	2.7	$H_{2.7} = 5634 + 70.2[BDPC] \times 10^5$	150–650	0.914
BDTC	2.5	$H_{2.5} = 3121 + 35.8[BDTC] \times 10^5$	125–500	0.996
LPC	3.8	$H_{3.8} = -11779 + 413.2[LPC] \times 10^5$	50–500	0.994
LTC	4.3	$H_{4.3} = 841 + 72.3[LTC] \times 10^5$	30–500	0.982
WPU	11.5	$H_{11.5} = 716 + 63129[WPU] \times 10^5$	0.15–5	0.914
WTU	14.0	$H_{14.0} = -261 + 16445[WTU] \times 10^5$	0.3–5	0.999
TPI	20.0	$H_{20.0} = -2417 + 427.7[TPI] \times 10^5$	15–500	0.990
TTI	8.7	$H_{8.7} = -4680 + 687.4[TTI] \times 10^5$	30–500	0.968
DBAPU	6.5	$H_{6.5} = -14861 + 4937[DBAPU] \times 10^5$	15–125	0.980
DBATU	6.5	$H_{6.5} = 4738 + 4004[DBATU] \times 10^5$	1–60	0.996

ple. It is possible that the injection volume was not the same for each run. Thus, all the peak heights were normalized by multiplying with a factor in the calculation.

The concentration of each species can be calculated by the following equation:

$$[\text{Compound}] = \frac{(H - C) \cdot df}{S} \cdot \frac{1}{100} \cdot \frac{1}{E}$$

where

[Compound] = concentration in N/l
H = normalized peak height, μV
C = intercept of the equations in Table 2
df = dilution factor, in g/g
S = slope of the equations in Table 2
E = equivalent weight of the compound

The products from phenyl isocyanate and *o*-tolyl isocyanate were not completely separated under the conditions for HPLC analysis. Since the ratio of phenyl and *o*-tolyl isocyanates added was 8/2, the mixture was calculated by multiplying the measured concentration with their respective fractions. For example, the concentration of *n*-butylcarbamate (BC) was calculated by the following equation:

$$[\text{BC}] = 0.8 \cdot [\text{BPC}] + 0.2 \cdot [\text{BTC}]$$

Selectivity of Isocyanate Reactions in OH-NCO Reactions under Various Conditions

1. URETHANE FORMATION

Reaction of n-*Butanol and Isocyanate in the Presence of Dabco T*

Five peaks appeared in the HPLC chromatograms of the solutions. The peaks with retention times of 2.8, 6.2, and 11 minutes were identified as the products of isocyanates with *n*-butanol (BC), dibutylamine (DBAU), and water (WU), respectively. As the reaction continued, the concentration of BC increased and that of DBAU decreased. Even with the precaution of passing dry nitrogen through the reaction flask, a trace amount of urea was detected. The profile of the reaction is shown in Figure 6.

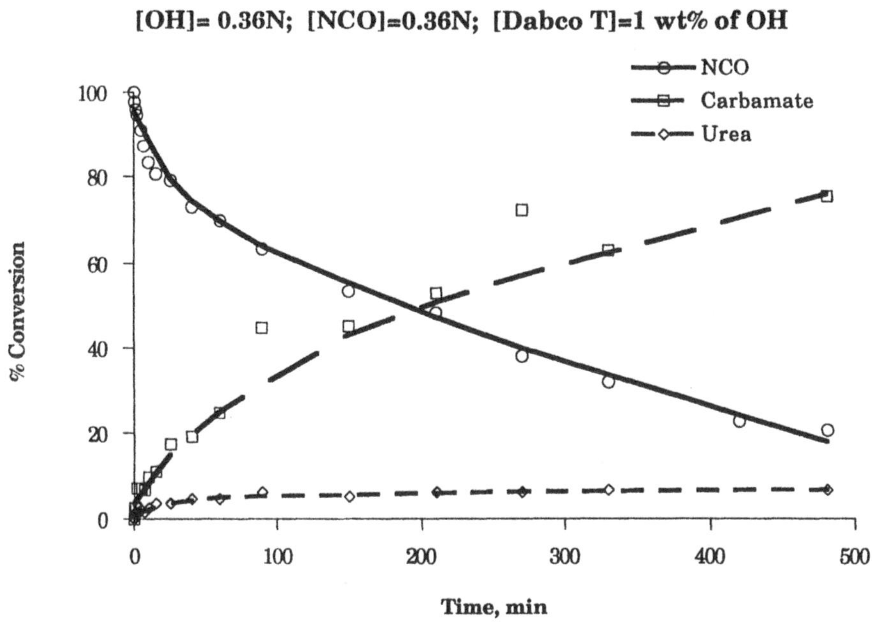

Figure 6. Selectivity of isocyanate reactions in n*-butanol and phenyl/o-tolyl isocyanate (80/20) catalyzed with Dabco T.*

No trimer was observed in this reaction even in the presence of catalyst Dabco T. We have run the same reaction in acetonitrile and observed trimer formation at the expense of urethane formation [22]. The solvent diglyme is not as polar as acetonitrile. This might account for the absence of trimer in the solution. Listemann et al. [23] also did not observe any trimer in the amine catalyzed reaction of *n*-butanol and phenyl/*o*-tolyl isocyanate in diglyme. Two peaks (retention times of 5.3 and 7.4 minutes) appeared in the chromatograms with almost the same peak height throughout the reaction. They were not identified. It is very likely these species are intermediates in the reaction. The sum of all the species equaled the initial concentration. It can be concluded that the concentrations of these intermediates must be very small.

Reaction of DB Oil and Isocyanate in the Presence of Dabco T

At low concentration, the chromatograms did not show any peaks corresponding to the products of DB oil and isocyanates. This may be attributed to the low extinction coefficient of the carbamates of DB oil. A second set of solutions three times as concentrated as the first set was analyzed by HPLC. Due to the high concentration of the solution, the peak corresponding to the DBAU was very broad and covered the peaks of two intermediates.

The concentration of carbamate based on DB oil (DBC) was calculated using the peak height at the retention time of 2.5–2.7 minutes. The reaction profile (Figure 7) of the DB oil and isocyanates is similar to that of *n*-butanol and isocyanates, only the rate of urethane formation was slower. No trimer was detected in this reaction. The concentration of urea in this reaction was higher than that in the *n*-butanol/isocyanate reaction.

Reaction of Lesquerella Oil and Isocyanate in the Presence of Dabco T

The chromatograms of the lesquerella oil and isocyanate reaction were much more complicated than those of the other two reactions. Again, the more concentrated solutions were used for the analysis due to the low extinction coefficient of carbamates based on lesquerella oil and isocyanates. In the calibration, it was noted that the peak height at retention time of 2.0 increased with the concentration of carbamate

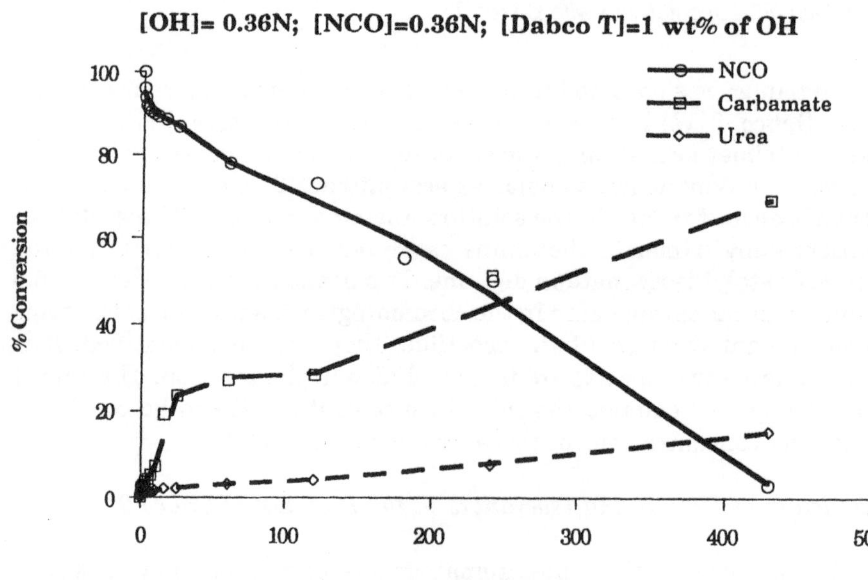

Figure 7. Selectivity of isocyanate reactions in DB oil and phenyl/o-tolyl isocyanate (80/20) catalyzed with Dabco T.

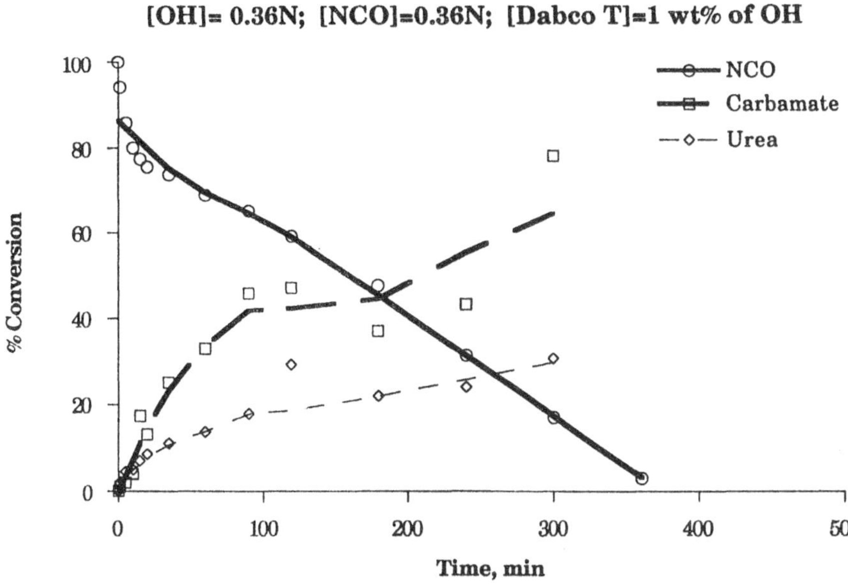

Figure 8. Selectivity of isocyanate reactions in lesquerella oil and phenyl/o-tolyl isocyanate (80/20) catalyzed with Dabco T in diglyme at 50°C.

based on lesquerella oil (LC). A least square line can fit the data points with very good correlation. However, the concentration of LC calculated from the peak height at retention time of 2.0 minutes was too high. The peaks at 3.8 and 4.3 minutes were used to do the calculation. The reaction profile is shown in Figure 8. In the beginning of the reaction, the formation of LC occurred at a faster rate than that of BC, but leveled off as the reaction continued.

No trimer was observed in this reaction. The urea content was even higher than that in the reaction of DB oil and isocyanate indicating less refinement of the lesquerella oil.

In the reactions of isocyanate and OH compounds catalyzed with Dabco T in diglyme, only urethane (carbamate) was detected. Figure 9 shows the comparison of urethane formation in the catalyzed reactions of three hydroxyl components. The rate of urethane formation in lesquerella oil and DB oil was faster than that of *n*-butanol at the beginning of the reaction, but leveled off later. The conversion of urethane was the lowest in the reaction of DB oil. Products through isocyanate crosslinking reactions such as allophanate, biuret, and isocyanurate trimer were not found. This is desirable for applications which require high elongation.

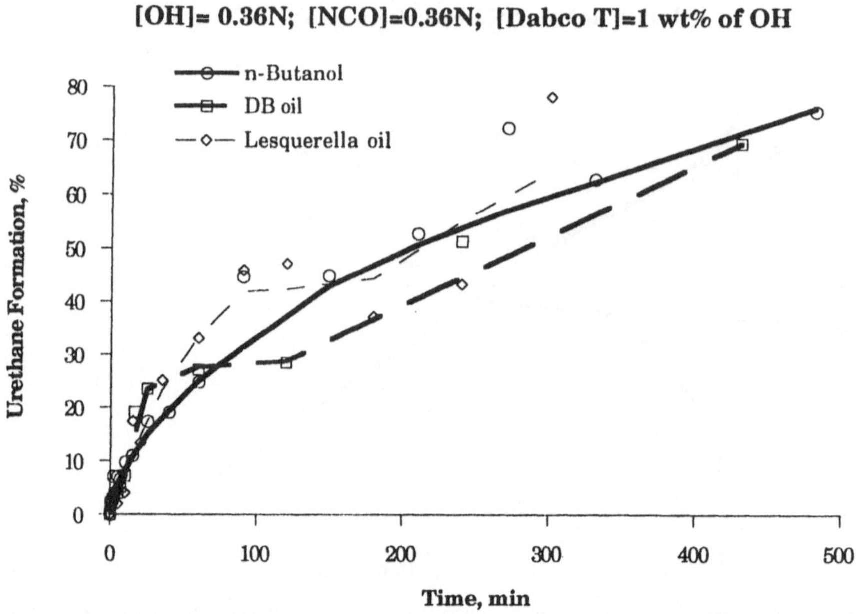

Figure 9. Urethane formation in reactions of hydroxyl compounds and phenyl/o-tolyl isocyanate (80/20) in diglyme at 50°C catalyzed with Dabco T.

2. URETHANE-ISOCYANURATE FORMATION

Reaction of n-*Butanol and Excess Isocyanate in the Presence of Dabco T and Dabco TMR-3*

The analysis of the reaction of *n*-butanol and excess isocyanate in the presence of Dabco T and Dabco TMR-3 is shown in Figure 10. Under the influence of a powerful trimerization catalyst (Dabco TMR-3), many peaks were detected in the chromatograms. However, due to the lack of standard compounds, they were not identified. Only the concentrations of carbamate (BC), urea from dibutylamine (DBAU), water and phenyl isocyanate (WPU), and trimer (TI) were calculated. The formation rate of urethane was much faster than that of trimer at the beginning of the reaction. The concentration of urethane and trimer was about the same at the end of the reaction. The discrepancy between the formation of urethane and trimer and the disappearance of NCO (about 20%) is due to the urea formation and other intermediate species which were not determined.

Reaction of DB Oil and Excess Isocyanate in the Presence of Dabco T and Dabco TMR-3

The reaction profile of DB oil reacting with excess isocyanate in the presence of Dabco T and Dabco TRM-3 is shown in Figure 11. Again the chromatograms of the quenched reaction solutions were quite complicated. Due to the overlapping of the peaks, only the concentrations of carbamates of DB oil, trimers and ureas were calculated. The results indicate that the rate of trimer formation was about the same as that of urethane formation.

Reaction of Lesquerella Oil and Excess Isocyanate in the Presence of Dabco T and Dabco TMR-3

The reaction of lesquerella oil and excess isocyanate in the presence of Dabco T and Dabco TMR-3 is shown in Figure 12. The initial rates of trimer and urethane formation were about the same. However, the concentration of trimer was about twice that of urethane at the end of the reaction.

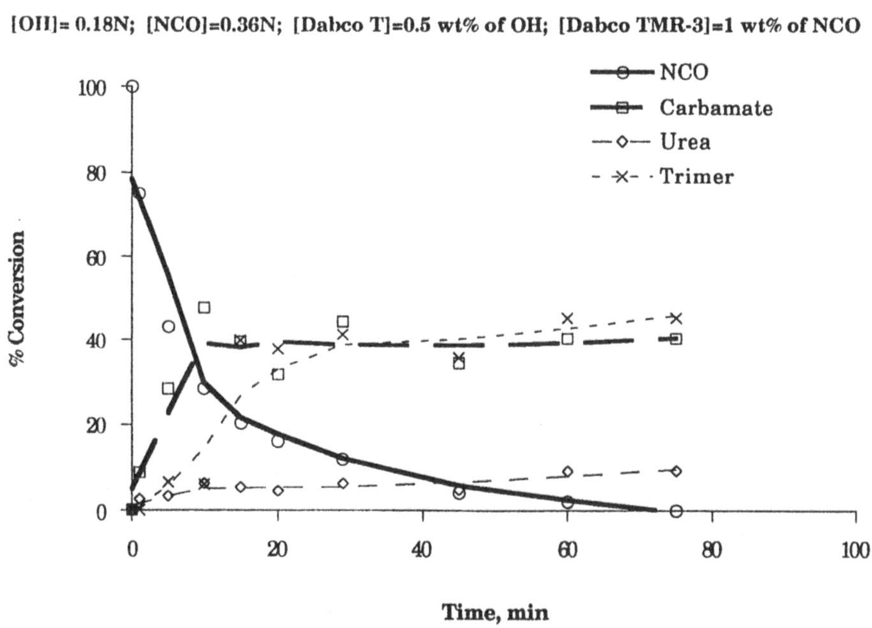

Figure 10. Selectivity of isocyanate reactions in n-butanol and phenyl/o-tolyl isocyanate (80/20) in diglyme at 50°C catalyzed with Dabco T and Dabco TMR-3.

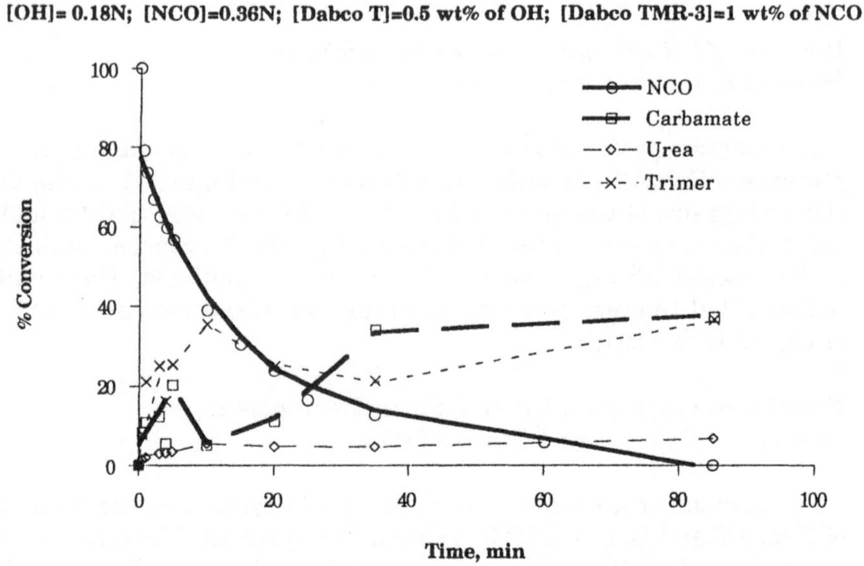

Figure 11. Selectivity of isocyanate reactions in DB oil and phenyl/o-tolyl isocyanate (80/20) in diglyme at 50°C catalyzed with Dabco T and Dabco TMR-3.

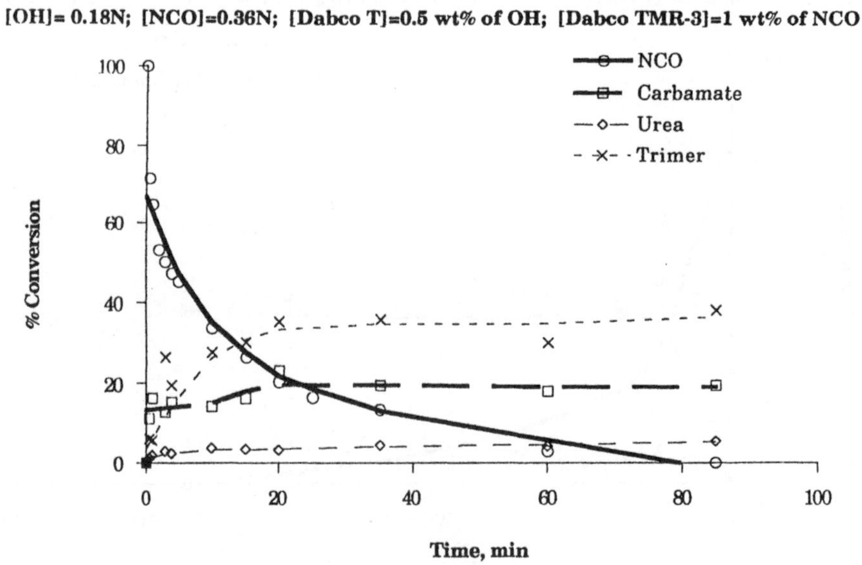

Figure 12. Selectivity of isocyanate reactions in lesquerella oil and phenyl/o-tolyl isocyanate (80/20) in diglyme at 50°C catalyzed with Dabco T and Dabco TMR-3.

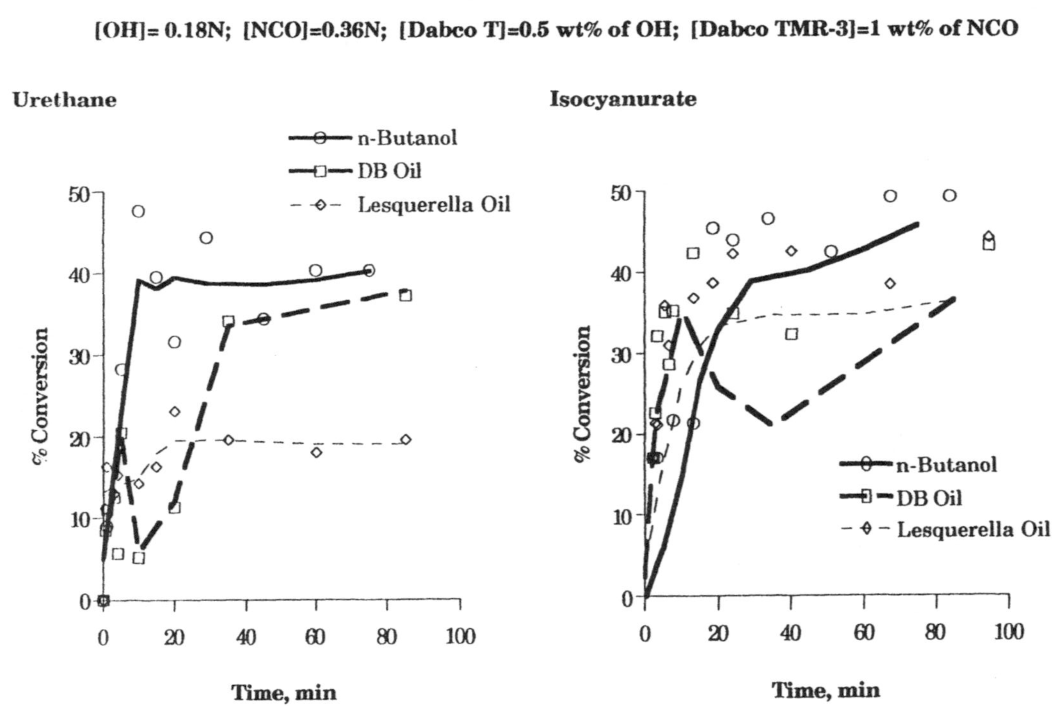

Figure 13. Urethane and isocyanurate formation in reactions of hydroxyl compounds and phenyl/o-tolyl isocyanate (80/20) in diglyme at 50°C catalyzed with Dabco T and Dabco TMR-3.

Figure 13 shows the comparisons of urethane and isocyanurate formation in the reaction of excess isocyanate with OH compounds catalyzed with both Dabco T and Dabco TMR-3. The concentration of Dabco TMR-3 was the same for all three reactions, but the concentration of Dabco T was based on the weight of the OH compounds. Under these conditions, the initial trimerization rate in the reactions of DB oil and lesquerella oil was faster than that of *n*-butanol. However, more isocyanurate was formed in the *n*-butanol reaction at the end when all the NCO had been consumed. On the other hand, the urethane formation in the lesquerella oil reaction was only about half as much as that in the DB oil and *n*-butanol reactions. The lesquerella oil actually encourages the isocyanurate formation.

Synthesis and Properties of Elastomers Based on Vegetable Oils

Polyurethane and polyurethane-isocyanurate elastomers based on castor and lesquerella oils were prepared using a one shot process. The materials used are listed in Table 3. These elastomers used castor oil or lesquerella oil as the soft segment, and an isocyanate and 1,4-butanediol or isocyanurate linkages as the hard segment. A carbodiimide modified methylene diphenylisocyanate, Isonate 143L, was mixed well with the hydroxyl component at room temperature. The mixture was then cured at 110 °C in a Carver hydraulic press. Catalyst Dabco T was used for the preparation of polyurethanes. Besides Dabco T, Dabco TMR-3 was added to accelerate the isocyanurate formation in the preparation of polyisocyanurate-urethane elastomers. These elastomers were cured at 110 °C for one hour on the press. Tables 4 and 5 show the composition and mechanical properties of these elastomers.

Polyurethane Elastomers

Castor oil has about 90% ricinoleic ester in its composition, which is the highest among all the vegetable oils [1]. However, due to the high molecular weight of ricinoleic ester, the resulting elastomers shown in Table 4 (PUC1 and PUC2) exhibited much lower properties than the elastomers PUP and PUT, which are based on commercial polyols derived from petroleum chemicals [24]. The soft segment of the polyether-based elastomer, PUP, was poly(oxytetramethylene) glycol (PTMEG) while that of the polyester-based PUT was poly(ϵ-caprolactone) diol (TONE polyol). The hard segment of PUP and PUT elastomers was composed of methylene-4,4′-di(phenyl isocyanate) (MDI) and 1,4-butanediol (BD). The three elastomers (PUC1, PUP, and PUT) has

Table 3. Materials.

Designation	Structure	Eq. Wt.
Lesquerella oil	56.3 wt% triglyceride of Lesquerolic fatty acid $CH_3\text{-}(CH_2)_5\text{-}CH(OH)\text{-}CH_2\text{-}CH{=}CH\text{-}(CH_2)_9\text{-}C(=O)\text{-}OH$	524
Castor oil (DB oil)	89.5 wt% triglyceride of Ricinoleic fatty acid $CH_3\text{-}(CH_2)_5\text{-}CH(OH)\text{-}CH_2\text{-}CH{=}CH\text{-}(CH_2)_7\text{-}C(=O)\text{-}OH$	342
PTMEG polyol	Poly(oxytetramethylene) glycol $HO\text{–}(CH_2CH_2CH_2CH_2O)_{13.6}\text{–}OH$	500
TONE polyol	Poly(caprolactone) polyol $H\text{-}[OCH_2CH_2CH_2CH_2CH_2C(=O)]_{10.8}\text{–}OH$	625
1,4-BD	1,4-Butanediol $HO\text{–}CH_2CH_2CH_2CH_2\text{–}OH$	45
MDI	4,4'-Methylene di(phenylisocyanate) 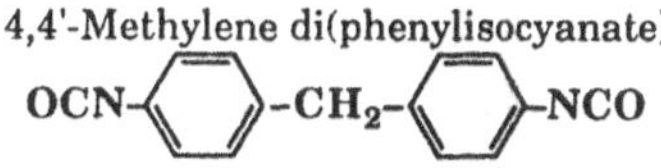	125
Isonate 143L	Carbodiimide-modified MDI	143

about the same soft segment content. The differences in their properties are apparent. The tensile strength of the castor oil–based elastomer (PUC2) could be made higher by increasing the hard segment (BD-MDI) content, but its elongation decreased drastically. This is due to the high crosslinks caused by the trifunctional castor oil. The elastomers based on castor oil and polyether polyol were not biodegradable. No bacteria or fungi growth was detected on these materials after 3 weeks incubation. On the other hand, the elastomers based on TONE polyol and lesquerella oil were biodegradable. A detailed description is discussed later.

Table 4 also shows polyurethane elastomers prepared with lesquerella oil as the soft segment and Isonate 143L and 1,4-BD as the hard segment. The mixture of lesquerella oil and Isonate 143L (PUL1) remained sticky after curing at 110°C overnight. This may be attributed to the other non-reactive ingredients in lesquerella oil acting as a plasticizer. The lesquerellic ester content in the oil is only slightly higher than 50% [25]. The lesquerella oil–based elastomer (PUL2)

Table 4. Polyurethane elastomers based on polyether and polyester polyols and polyols derived from vegetable oils.

Designation	PUP	PUT	PUC1	PUC2	PUL1	PUL2	PUL3	PULC
Formulation, phr								
PTMEG 1000	62.89	—	—	—	—	—	—	—
TONE 1250	—	67.93	—	—	—	—	—	—
DB Oil	—	—	67.62	51.81	—	—	—	12.56
Lesquerella Oil	—	—	—	—	75.76	57.18	44.75	62.79
1,4-Butanediol	5.66	4.89	—	5.18	—	5.72	8.95	—
Dabco T-12	0.05	0.05	—	—	—	—	—	—
Dabco T	—	—	1.35	1.17	1.50	0.06	0.05	0.08
MDI	31.45	27.17	—	—	—	—	—	—
Isonate 143L	—	—	31.02	41.84	22.73	37.04	46.25	24.57
Isocyanate Index	100	100	100	102	98	105	105	105
Physical Properties								
100% Modulus, psi (MPa)	800 (5.52)	750 (5.17)	418 (2.88)	—	sticky	489 (3.37)	—	74 (0.51)
Tensile Strength, psi (MPa)	4446 (30.66)	6667 (45.98)	1300 (8.97)	2476 (17.08)		781 (5.39)	2294 (15.82)	196 (1.35)
Elongation, %	673	570	267	64		183	105	323
Biodegradability								
ASTM G22, viable cell count after 12 days, CFU/ml	7.6×10^6	3.8×10^8		8.5×10^6		3.8×10^8	—	—
ASTM G22, after 21 days	no growth	+2 to +3/+4		no growth		+3 to +4/+4	+2 to +3/+4	+3 to +4/+4

Table 5. Polyisocyanurate-urethane elastomers based on polyols derived from vegetable oils.

Designation	PUIC	PUIL1	PUIL2	PUIL3
Formulation, phr				
DB Oil	52.54	—	—	—
Lesquerella Oil	—	62.68	62.11	48.84
1,4-Butanediol	—	—	—	4.88
Dabco T	1.05	1.25	2.48	1.10
Dabco TMR-3	0.43	0.34	—	—
MDI	—	—	—	—
Isonate 143L	45.97	35.73	35.40	45.18
Isocyanate Index	191	186	176	144
Physical Properties				
10% Modulus, psi (MPa)	—	—	—	—
Tensile Strength, psi (MPa)	4117 (28.39)	915 (6.31)	923 (6.37)	2790 (19.24)
Elongation, %	10	66	68	49
Biodegradability				
ASTM G22, viable cell count after 12 days, CFU/ml	8.5×10^6	1.1×10^9	—	—
ASTM G21, after 21 days	no growth	+3 to +4/+4	+3 to +4/+4	+3 to +4/+4

which had about the same equivalent ratio as did castor oil (PUC2) showed much poorer properties. By increasing the hard segment content, the lesquerella oil–based elastomer (PUL3) showed comparable stress-strain properties to those of PUC2 but with a slightly higher elongation. The advantage of PUL3 is that this elastomer was biodegradable, although it was not as accessible as the one containing less hard segment.

One elastomer (PULC) was prepared using a mixture of lesquerella oil and castor oil and the soft segment. This elastomer was biodegradable, although its properties were very poor. Incorporation of short chain diol in the hard segment might boost its properties. However, the increasing hard segment content might also affect the biodegradability of the material.

Polyisocyanurate-Urethane Elastomers

Table 5 shows the polyisocyanurate-urethane elastomers based on castor oil and lesquerella oil. The isocyanurate linkages were introduced in the polymer network by using high isocyanate index and the trimerization catalyst Dabco TMR-3 (catalyze the formation of isocyanurate). The tensile strength of the castor oil–based elastomer containing isocyanurate (PUIC) could be made comparable to that of polyether urethane elastomer PUP. However, the elongation of this elastomer was only 10%. Elastomer PUIC was also not biodegradable.

Three isocyanurate-containing urethane elastomers (PUIL1–PUIL3) were prepared using lesquerella oil as the soft segment. In the elastomer PUIL3, the hard segment included chain extender 1,4-BD as well as isocyanurate linkages. The isocyanate index in PUIL3 was lower than that in the other two. The tensile strength of elastomer PUIL3 was much better than those of elastomers PUIL1 and PUIL2. Due to the presence of non-reactive ingredients in lesquerella oil, the elongation of these samples remained about the same. It should be noted that the biodegradability of the lesquerella oil–based polymers depends on the content of lesquerella oil, regardless of the type of hard segment. The lesquerella oil content in elastomer PUIL1 is about 63 wt% while that of elastomer PUL2 was 57 wt%. The bacteria cell count in PUIL1 is higher than that in PUL2, even though the hard segment in PUIL1 is isocyanurate and that in PUL2 is urethane.

Thermal Properties of Polymers

Thermal Mechanical Analysis

Figures 14 and 15 show the TMA scans of elastomers based on les-

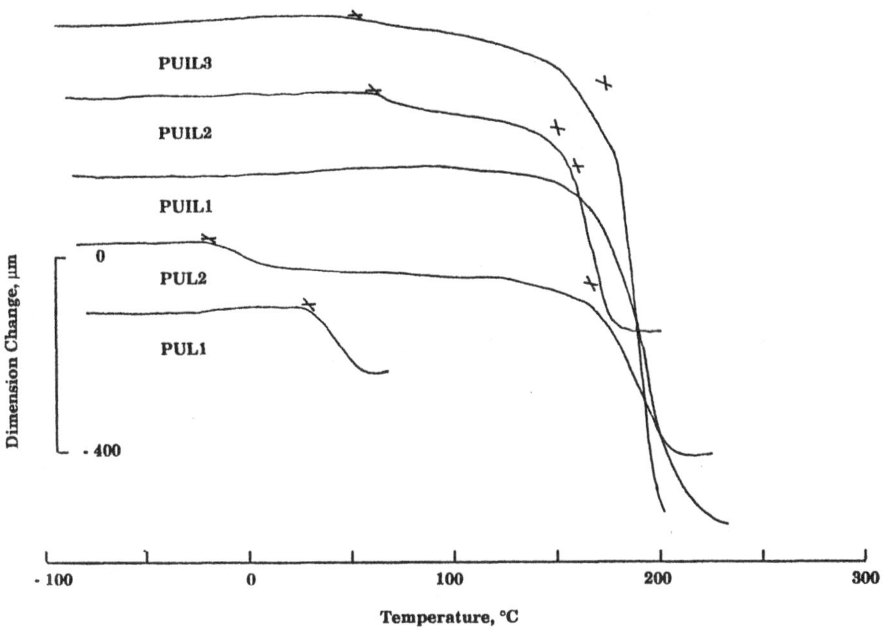

Figure 14. TMA thermograms of lesquerella oil-based elastomers.

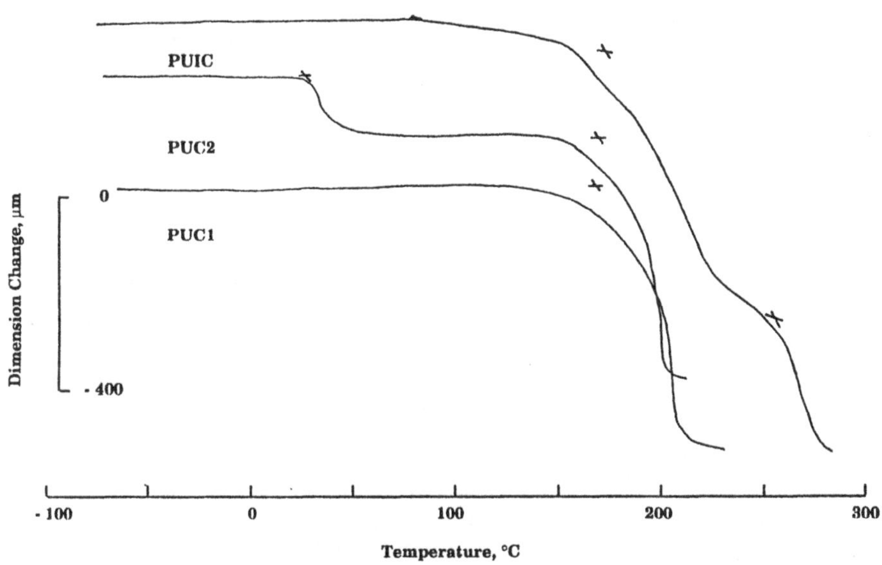

Figure 15. TMA thermograms of castor oil-based elastomers.

querella oil and castor oil, respectively. The transition was defined from the derivative of the TMA curve (dimension change versus temperature) and the onset temperature was used as the transition temperature of the elastomer. The thermal transitions of these elastomers are listed in Table 6.

The elastomer PUL1, which contained only isocyanate and lesquerella oil, softened at 26 °C. This explains why the material remained sticky after curing. When a chain extender was incorporated into the system, elastomer PUL2, its softening temperature increased close to 170 °C. A glass transition was also detected at −25 °C. The softening temperature of three isocyanurate-containing elastomers, PUIL1–PUIL3, was about the same as that of elastomer PUL2. Their glass transition temperature shifted to 55 °C. This can be attributed to the highly crosslinked network which was caused from the trifunctional lesquerella oil and isocyanurate. However, due to the presence of nonreactive components acting as a plasticizer, the softening temperature of these urethane-isocyanurate elastomers remained the same as that of urethane elastomer.

The castor oil–based polyurethane elastomers exhibited a softening temperature comparable to that of lesquerella oil. The one which was chain extended with 1,4-butanediol showed a glass transition at 24 °C. The isocyanurate-containing elastomer, PUIC, showed a second softening point at a much higher temperature. This is probably due to the isocyanurate linkages present in the system.

Differential Scanning Calorimetry

The DSC scans of the vegetable oil–based elastomers are shown in Figures 16 and 17. The glass transition temperature of these materials was determined by the midpoint method. In some cases, an onset temperature was used on the curves which showed no apparent glass transition but did exhibit a slope change. All the materials began to show an endotherm or an exotherm at temperatures of about 200 °C. This may indicate the occurrence of reactions of double bonds present in the vegetable oil and that of the unreacted isocyanates.

The lesquerella oil–based polyurethane elastomer PUL1 exhibited only a soft segment glass transition temperature at −34 °C (Table 6). The glass transition temperature of elastomer PUL2, which contained MDI-BD hard segment, shifted to −21 °C. In addition, a second transition was observed at 65 °C. This transition can be assigned as the dissociation of short range ordered hard segment. The same effect was also observed in the polyurethane elastomers based on castor oil, only their

Table 6. Thermal properties of elastomers based on vegetable oils.

	Thermal Transitions				Thermal Stability, TGA		
	DSC		TMA		Wt. Loss Temperature, °C		
Sample	T_g, °C Midpoint	T_g, °C Onset	T_g, °C Onset	T_s, °C Onset	2%	10%	Residual Wt. at 450°C, %
PUL1	−34.0			26.1			
PUL2	−21.4	65.1	−25.0 68.7	167.8			
PUIL1	—	78.1		161.3	207.5	336.0	38.7
PUIL2	−32.2	79.8	57.2	151.6	187.5	325.0	41.7
PUIL3	—	76.7	55.8 109.6	180.0	194.5	315.5	35.6
PUC1	9.6	—	—	168.9	186.0	315.0	18.4
PUC2	19.8	—	23.5	171.9	256.5	315.5	24.6
PUIC	—	62.8	72.3	174.7 259.1	218.0	334.0	44.0

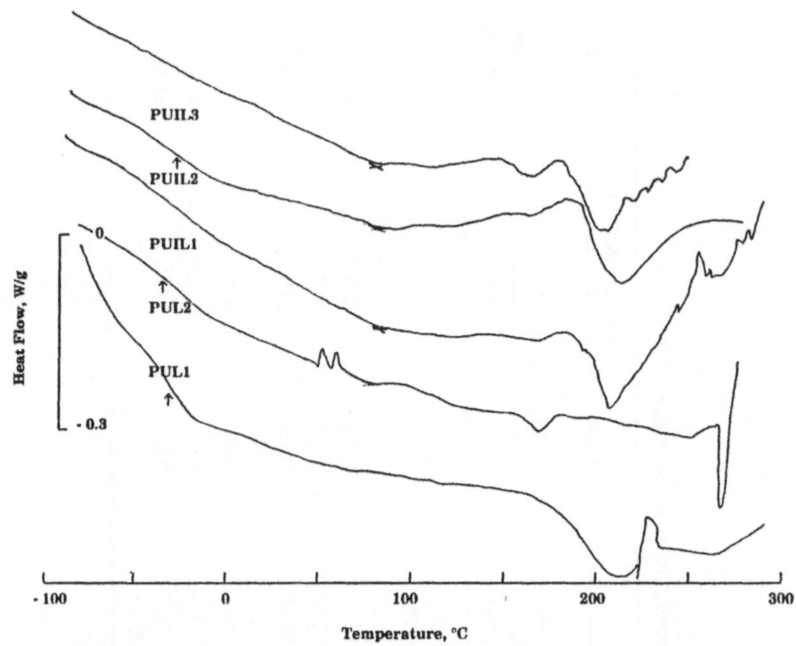

Figure 16. *DSC thermograms of lesquerella oil-based elastomers.*

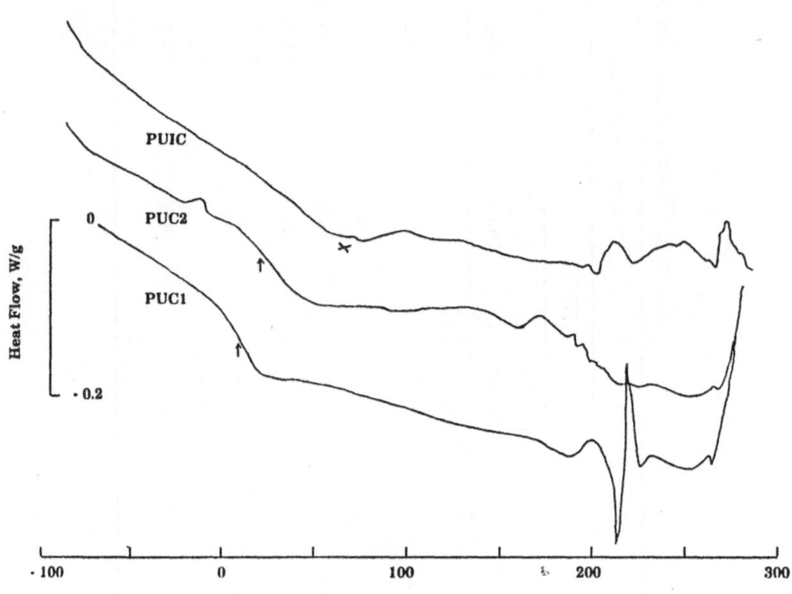

Figure 17. *DSC thermograms of castor oil-based elastomers.*

soft segment glass transition temperatures were higher (i.e., 10 and 20°C) due to the low molecular weight of DB oil. No dissociation of short range ordered hard segment was observed in the castor oil–based polyurethane elastomers.

All the isocyanurate-containing elastomers exhibited dissociation of short range ordered hard segment in temperatures ranging from 63° to 80°C. The temperature for elastomers based on lesquerella oil was higher than that of castor oil. Only elastomer PUIL2 showed a soft segment glass transition at −32°C.

The DSC is a much more sensitive instrument than TMA for measuring the glass transition temperature of elastomers, especially the soft segment glass transitions. As shown in Table 6, the temperatures of transitions which were measured by the two methods are quite close. The results indicate that there was a large amount of crosslinking and phase mixing of soft and hard segments in these elastomers.

Thermogravimetric Analysis

The results from TGA can be used to rank the thermal stability of polymers. Figures 18 and 19 show the TGA scans of lesquerella oil–

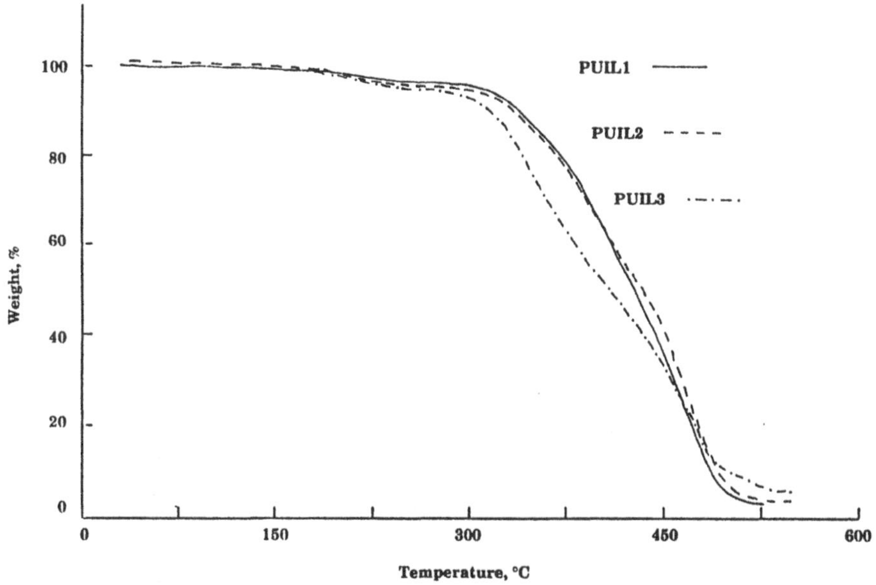

Figure 18. *TGA thermograms of lesquerella oil–based elastomers.*

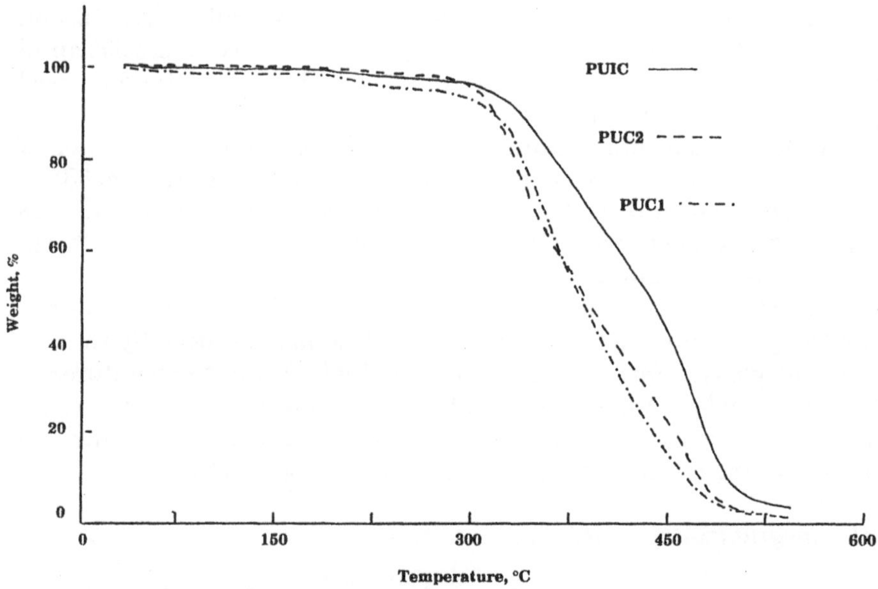

Figure 19. TGA thermograms of castor oil–based elastomers.

and castor oil–based elastomers. The temperatures at which 2% and 10% weight loss occurred and the residual weight at 450 °C are summarized in Table 6. The results indicate that the initial temperature of weight loss for the vegetable oil–based elastomers was much lower than those based on petrochemicals. The results from a previous study in our laboratory indicate that the temperature at 2% weight loss for a polyether urethane composed of MDI-PPG-BD was about 100 °C higher than that of urethane and urethane-isocyanurate elastomers based on vegetable oils [26]. However, the residual weight at 450 °C of the PPG-based elastomers was only 22% rather than 40% for the vegetable oil–based urethane-isocyanurate elastomers. The low initial temperature for the vegetable oil–based elastomers can be attributed to the non-reactive ingredients present in the vegetable oils. The effect of isocyanurate linkage on the thermal stability of these elastomers is clearly demonstrated from the residual weight of the castor oil–based elastomers.

BIODEGRADABILITY

Biodegradability testing by the ASTM G22 method which uses the bacterium *P. aeruginosa* ATCC 13388, was carried out. Three trial isocyanate-based polymer samples were examined by the technique. The last of the three tested appeared to demonstrate a trace reaction (+1 out of +4 total) for biodegradability. Small colonies grew around and on the polymer sample square.

ASTM G22 and G21 tests permit the incubation of many test samples at the same time and over a long period but the scoring of results is not quantitative in nature. Therefore, comparing biodegradability of one polymer with another is difficult.

The polymer samples based on vegetable oils and other polyols were examined by ASTM G21 which uses five ATCC strains of fungi including *Aspergillus niger* (ATCC 9642), *Penicillium funiculosum* (ATCC 9644), *Chaetomium globosum* (ATCC 6205), *Gliocladium virens* (ATCC 9645), and *Aureobasidium pullulans* (ATCC 9348). Growth of fungi completely covered a filter paper square that served as a positive control (rating of +4). Polymers based on castor oil and polyether polyol showed no growth after incubation for 21 days. The elastomer made from TONE polyol showed a growth rating of +2 to +3 out of +4. Polymer PUL2 (lesquerella oil) produced a growth rating of +3 to +4 out of +4 total while polymer PUL3 showed a growth rating of +2 to +3. Polymer PUCL (blend of castor oil and lesquerella oil) demonstrated a growth rating of +3 to +4.

An attempt was also made to create a quantitative broth assay based on ASTM G22. Instead of petri plates, 300 ml flasks were used as containers. The salt's broth was prepared according to ASTM G22 except the agar (hardening agent) was not added. Once inoculated, flasks were incubated on a shaker water bath at 37°C. Growth of bacteria was followed by measuring culture optical density (OD) on a Klett-Summerson colorimeter and the standard colony count assay.

Polymer samples PUP (PTMEG), PUT (TONE), PUC2 and PUIC (castor oil) and PUL2 and PUIL1 (lesquerella oil) were tested quantitatively using the ASTM G22 procedure with some modification. Flasks containing salt's broth were used which were incubated on a shaker water bath. Growth of bacteria was followed by culture optical density and the standard colony count assay. A flask containing *P. aeruginosa* ATCC 13388, polymer PUIL1 and salt's broth demonstrated an increase in bacterial numbers from 1.8×10^5 cells/ml to 1×10^9 cells/ml after 12 days' incubation (Figure 20) whereas the control showed a bacterial growth from 2×10^5 cells/ml to 4×10^6 cells/ml after 21 days' incubation. The growing curves of elastomers based on castor oil and PTMEG were about the same as that of the control. The bacteria grew in the medium containing elastomer from TONE polyol, but the cell

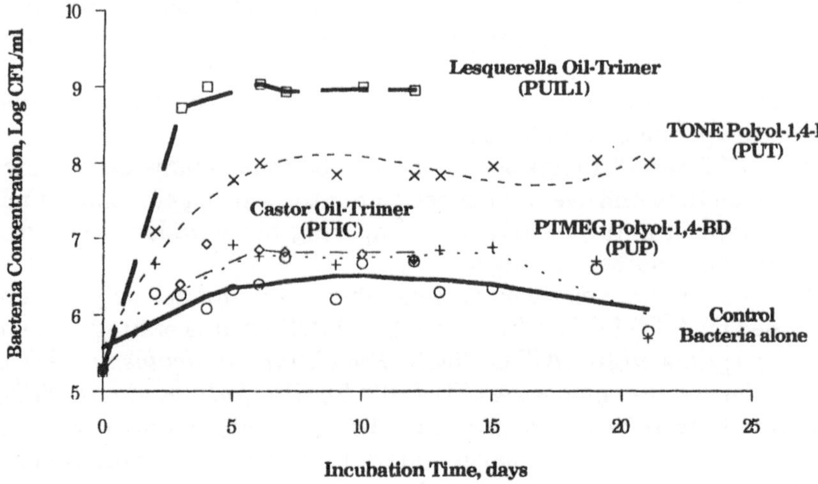

Figure 20. Bacterial growth in polyurethane elastomers based on various polyols.

count was not as high as that in the lesquerella oil–based elastomer.

It is surprising to find that the elastomers based on lesquerella oil were biodegradable but those based on castor oil were not under the same test conditions. The differences between the structure of ricinoleic acid (castor oil) and lesquerella acid (lesquerella oil) are an ethylene group on their hydrocarbon backbones. The content of fatty acids of oleic, linolenic and linoleic in lesquerella is much higher than that in castor oil (Table 7). It is essential to study the degradability behavior of each fatty acid to find out which one is responsible. Further study should also be made using different strains of bacteria on the materials based on lesquerella oil.

CONCLUSIONS

The reactivity of lesquerella oil, castor oil and *n*-butanol with isocyanates was investigated by model reactions in diglyme solutions in the presence of catalysts and various reactants such as water and excess isocyanate to simulate actual polymerization processes. A mixture containing 80% phenyl isocyanate and 20% *o*-tolyl isocyanate was used for the reactions. By varying the amount of catalysts, the rate of reaction could be made comparable for these three compounds.

The selectivity of various isocyanate reactions was measured using high performance liquid chromatography (HPLC). No trimer was detected in the reaction of OH and NCO catalyzed with Dabco T. In the reaction of OH and excess NCO catalyzed with Dabco T and Dabco TMR-3, the rate of carbamate and trimer formation was about the same with *n*-butanol and castor oil. More trimer was observed in the reaction of lesquerella oil.

Table 7. Composition of fatty acids in vegetable oils.

Fatty Acid	Structure	Content, %
	Lesquerella Oil	
Lesquerella	$CH_3\text{-}(CH_2)_5\text{-}CH(OH)\text{-}CH_2\text{-}CH{=}CH\text{-}(CH_2)_9\text{-}C(=O)\text{-}OH$	56.3
Oleic	$CH_3(CH_2)_7CH{=}CH(CH_2)_7COOH$	17.7
Linolenic	$CH_3\text{-}[CH_2\text{-}CH{=}CH]_3\text{-}(CH_2)_7COOH$	12.7
Linoleic	$CH_3(CH_2)_4\text{-}CH{=}CH\text{-}CH_2\text{-}CH{=}CH\text{-}(CH_2)_7COOH$	7.6
Stearic	$CH_3(CH_2)_{16}COOH$	1.7
Palmitic	$CH_3(CH_2)_{14}COOH$	1.3
Misc.		2.7
	Castor Oil	
Ricinoleic	$CH_3\text{-}(CH_2)_5\text{-}CH(OH)\text{-}CH_2\text{-}CH{=}CH\text{-}(CH_2)_7\text{-}C(=O)\text{-}OH$	89.5
Linoleic	$CH_3(CH_2)_4\text{-}CH{=}CH\text{-}CH_2\text{-}CH{=}CH\text{-}(CH_2)_7COOH$	4.2
Oleic	$CH_3(CH_2)_7CH{=}CH(CH_2)_7COOH$	3.0
Palmitic	$CH_3(CH_2)_{14}COOH$	1.0
Stearic	$CH_3(CH_2)_{16}COOH$	1.0
Dihydroxystearic	$(OH)_2C_{17}H_{33}COOH$	0.7
Linolenic	$CH_3\text{-}[CH_2\text{-}CH{=}CH]_3\text{-}(CH_2)_7COOH$	0.3
Eicosanoic	$CH_3(CH_2)_{18}COOH$	0.3

Polyurethane and polyurethane-isocyanurate elastomers based on castor and lesquerella oils were synthesized. Due to the presence of other non-reactive fatty acid esters, the properties of these elastomers were not as good as the ones based on standard urethane components. However, it was discovered that the elastomers based on lesquerella oil were biodegradable. The extent of degradation was inversely related to the hard segment content.

These special vegetable oils are expensive compared to the polyols derived from petro-chemicals. For example, lesquerella oil costs more than ten dollars a pound and is in short supply. The purpose of this study was to find some special properties from these renewable oils. Hopefully a new industry can be created from this work. Currently, the lesquerella plant grows wild in the southwestern desert in the United States. With the newly found applications, this vegetation can be cultivated as an economical crop in that area.

ACKNOWLEDGEMENTS

The authors gratefully acknowledge the University of Detroit Mercy's Center of Excellence in Environmental Engineering and Science and the U.S. Environmental Protection Agency for funding these studies through assistance agreement #CR820478-01-0.

REFERENCES

1. CasChem Technical Bulletin 100, biothane systems.
2. Frankel, E. N., F. L. Thomas and W. K. Rohwedder. 1973. *I & EC Prod. Res. Develop.*, 12:47.
3. Carlson, K. D., W. J. Schneider, S. P. Chang and L. H. Princen. 1981. Chapter 21 in *New Sources of Fats and Oils,* E. H. Pryde, L. H. Princen and K. D. Mukherjee, eds., Champaign, IL: American Oil Chemists' Society.
4. Carraher, C. E., Jr. and L. H. Sperling, eds. 1986. *Renewable-Resource Materials: New Polymer Sources.* New York: Plenum.
5. Sperling, L. H. and J. A. Manson. 1983. *JAOCS,* 60:1887.
6. Oertel, G. 1985. *Polyurethane Handbook, PU Based on Castor Oil Polyols in Telecommunications.* Hanser Publisher.
7. Cuadrado, T. R. and R. J. J. Williams. 1986. *Macromol. Chem., Macromol. Symp.,* 2:61–68.
8. Kumar, R., S. K. Srivastava and G. N. Mathur. 1985. *J. Elastomers & Plast.,* 17(1):89.

9. Kumar, R., S. K. Srivastava and G. N. Mathur. 1983. *Indian J. Technol.,* 21(12):515–518.
10. Ghatge, N. D. and V. B. Phadke. 1967. *J. Appl. Polym. Sci.,* 11(5):629.
11. Petrovic, Z. S. and D. Fajnik. 1984. *J. Appl. Polym. Sci.,* 29:1031.
12. Saxena, P. K., S. K. Menon and S. R. Srinivasan. 1986. *J. Polym. Mater.,* 3(4):251.
13. Knaub, P. and Y. Camberlin. 1986. *Eur. Polym. J.,* 22:633.
14. Khoe, T. H., F. H. Otey and E. N. Frankel. 1972. *JAOCS,* 49:615.
15. Lyon, C. K., V. H. Garrett and E. N. Frankel. 1974. *JAOCS,* 51:331.
16. Kase, M., N. Ogoshi and K. Tsuyusaki. 1987. JP 62,209,124.
17. Geng, K., Q. Chen, M. Gu and X. Yu. 1987. *Gaodeng Xuexiao Huaxue Xuebao,* 8(11):1044.
18. Darby, R. T. and A. M. Kaplan. 1968. *Appl. Microbiol.,* 6:900.
19. Huang, S. J., C. Macri, M. Roby, C. Benedict and J. A. Cameron. 1981. In *ACS Symp. Ser.,* K. N. Edwards, ed., 172:471.
20. Seal, K. J. and R. A. Pathirana. 1982. *Int. Biodeterior. Bull.,* 18:81.
21. Huang, S. J. 1985. "Biodegradable Polymers," in *Encyclopedia of Polym. Sci. & Eng., 2nd Edition,* H. F. Mark, N. M. Bikales, C. G. Overberger and G. Menges, eds., New York: John Wiley & Sons, 2:220–243.
22. Wong, S. and K. C. Frisch. 1986. *J. Polym. Sci., Part A: Polym. Chem.,* 24:2867–2875, 2877–2890.
23. Listemann, M. L., A. C. Savoca and A. L. Wressell. 1991. *Polyurethanes World Congress,* p. 524.
24. Kresta, J. E. and I. Horacek. Private communication.
25. Flora Lipids™ Technical Bulletin, "Specialty Vegetable Oils on Lesquerella Oil."
26. Frisch, K. C. and S.-W. Wong. 1990. "Isocyanurate-Containing Polyurethanes," *Cellular Polymers,* pp. 433–470.

p-Phenylene Diisocyanate Based Polyurethane Elastomers

AJAIB SINGH
Manager Research and Development
Adiprene®/Vibrathane®
Uniroyal Chemical Company, World Headquarters
Middlebury, CT 06749, U.S.A.

ABSTRACT

p-Phenylene diisocyanate (PPDI) has been well documented for its usefulness in providing high performance polyurethane thermoplastic and castable elastomers. The monomeric PPDI is solid below 95°C, yet its vapor pressures are similar to that of toluene diisocyanate. The tendency to sublime, however, makes PPDI somewhat more difficult to handle than the toluenediisocyanate. PPDI is somewhat more reactive than other commonly used aromatic diisocyanates. A large difference in the reactivity of the two isocyanate groups in PPDI makes it well-suited for making polyurethane prepolymers with relatively low levels of the residual monomer contents. PPDI can provide both thermoset and thermoplastic polyurethane elastomers with excellent mechanical and engineering properties. PPDI prepolymers are suitable for curing with both the diamine type as well as the diol type of curatives, providing very strong hard segments. These rigid hard segments are responsible for the unique combination of superior dynamic properties, excellent toughness, good hydrolytic stability, exceptional flex fatigue, and high abrasion resistance of its elastomers relative to other commercially available polyurethane elastomers.

para-PHENYLENE DIISOCYANATE (PPDI)

MONOMERIC PARA-PHENYLENE DIISOCYANATE (PPDI) HAS BEEN KNOWN since 1913 [1]. Many studies pertaining to the use of this diisocyanate in thermoplastic and castable polyurethane elastomers have been documented in the literature [2–19]. The monomeric *p*-phenylene diisocyanate was first commercialized by AKZO Chemie bv (Holland) and by ARMAK Company (Chicago, U.S.A.) in the early 1980s under the tradename Elate™ 160 [20], but the monomer and castable prepolymers derived from it were withdrawn from the market after a few years. The AKZO process for manufacture of PPDI monomer was based on a modified Hoffman reaction [34]. In the late eighties, E. I. du Pont de Nemours and Co. announced plans for the production and commercialization of this monomer. Their process for the production of PPDI is based on the phosgenation of *p*-phenylene diamine [15]. The monomer is now available through du Pont and is produced by a joint production arrangement between du Pont and Uniroyal Chemical Company [21].

PHYSICAL PROPERTIES OF *p*-PHENYLENE DIISOCYANATE

The monomeric PPDI is a white solid at room temperature and is available in flaked form. Its molecular weight is 160.13 with an NCO content of 52.5% and specific gravity of 1.170 at 100°C. The monomer melts at 94–95°C and has a boiling point of 260°C at 760 mm Hg (or 110–112°C at 25 mm Hg). It has a tendency to sublime even at temperatures as low as 80°C, making its handling somewhat difficult. The monomer is stable at room temperature when stored in sealed containers for up to 6 months (AKZO and ARMAK company literature). However, slow dimerization occurs at and above its melting point, making its storage in liquid form undesirable.

TOXICITY OF *p*-PHENYLENE DIISOCYANATE

According to the manufacturer's literature, PPDI is toxic by ingestion. Although it is non-toxic by dermal route, PPDI is a skin and eye irritant and skin and lung sensitizer. It is judged to be non-mutagenic under the Ames test. Although the vapor pressure of PPDI is very similar to that of toluene diisocyanate (TDI) [9], and it should minimally be handled with all the precautions and safeguards currently employed with TDI, its tendency to sublime makes PPDI somewhat more difficult to handle and may require additional care. Some additional toxicity information for PPDI is summarized in Table 1.

REACTIVITY OF *p*-PHENYLENE DIISOCYANATE

The reactivity of PPDI has been studied by several investigators [6,9,14,19,22,23,25]. In a reaction rate study with 2-ethylhexanol, it was found that both meta- and para-phenylene diisocyanates show higher

Table 1. Toxicity data. *

Toxicity Test	
Acute oral toxicity (LD50)	2160 mg/kg
Acute dermal toxicity (LD50)	4640 mg/kg
Acute inhalation toxicity (LC50)	10.96 mg/L (1 hour)**
Primary irritation index (skin)	1.13
Eye irritation score	21†

*AKZO Product Literature, Development Bulletin 2 ELate 160, dated August 1980 and Du Pont Chemicals MSDS for "Hylene" PPDI, 3034 CR dated June 1992.

**Tested as aerosol at saturation point at 27°C. Because PPDI has vapor pressure at 20°C of less than 0.02 mm Hg, it may be assumed to have a Threshhold Limit Value approximately equal to TDI or 0.02 ppm.

†Average score for 6 rabbits. Total score is 110; scores greater than 30 indicate potential to cause irreversible damage.

reactivity than the phenyl isocyanate. Both also show a significant decrease in slope of the rate curves as the reaction proceeds [22]. This decrease in the reaction rate with the extent of completion of reaction as seen in the data presented in Table 2, clearly shows that the presence of an isocyanate group on the benzene ring imparts approximately a 6 fold increase in the reactivity of the other isocyanate, the meta position being slightly more activating than the para in phenylene diisocyanate. As the reaction of phenylisocyanate proceeds, the urethane group formed exerts less activating influence than the isocyanate group.

The relative effects of various substituent groups on the reactivity of aromatic isocyanates were also studied by Burkus and Eckert [23] in

Table 2. First order reaction rates for some aromatic diisocyanates at various degrees of reaction.

	$k \times 10^4$ sec^{-1} at Extent of Reacted Isocyanate		
	10%	50%	90%
Phenylisocyanate	1.2	1.1	1.1
p-Phenylene diisocyanate	5.7	3.7	1.6
m-Phenylene diisocyanate	6.0	4.4	2.5
2,4-Toluene diisocyanate	2.0	1.2	0.22
2,6-Toluene diisocyanate	0.8	0.32	0.18
4,4-Methylene bis (phenylisocyanate)	1.7	1.3	0.9

Data Source: (Reference [22])—Bailey, Kirss, and Spaunburgh, *Industrial and Eng. Chem.*, 48, 794–797 (1956).

triethylamine catalyzed reactions of diisocyanates with 1-butanol in toluene at 39.7°C. These investigators were able to calculate the second order rate constants for each of the isocyanate group in several aromatic and aliphatic diisocyanates. The reactivity differences for the first and second isocyanate groups of several commercially important aromatic diisocyanates is summarized in Table 3.

The relative reaction rates for the first and second isocyanate groups are important in the production of polyurethane prepolymers as they control levels of residual (unreacted monomeric) diisocyanate in the prepolymers. The greater the reactivity difference between the two isocyanate groups in a diisocyanate, the lower is the expected extent of the residual monomeric diisocyanate in the prepolymer formed by reacting a molar excess of the diisocyanate with a polyol. Due to toxicological implications associated with handling and processing of polyurethane prepolymers, the lower levels of free diisocyanate content are highly desirable. Based on the k_1/k_2 data in Table 3, one would expect higher levels of free MDI, while the free TDI or PPDI levels are significantly lower (and similar) in their respective prepolymers. This indeed is true when equal molar ratios of the diisocyanates to polyol are fully reacted in producing the prepolymers. Since the levels of free monomer in TDI- and PPDI-based prepolymers, made under similar diisocyanate/polyol molar ratios are similar, and their vapor pressures under normal use temperatures are also similar, one must, at least, use similar cautions and safety practices in handling PPDI prepolymers as with TDI prepolymers. In practice, the need for special safety precautions (proper ventilation and disposal methods) is emphasized [13] while handling PPDI and PPDI-based prepolymers to fully protect workers from the potentially hazardous and sensitizing nature of this diisocyanate.

In a more recent study on the reactivity of PPDI and other commercial diisocyanates, Wong, Juang, Olgac and Frisch [6,7] observed

*Table 3. Reaction rate constants for first (k_1) and second (k_2) isocyanate group in different diisocyanates.**

Diisocyanate	k_1 l. mole^{-1} min^{-1}	k_2 l. mole^{-1} min^{-1}	k_1/k_2
p-Phenylene diisocyanate	3.15	0.342	9.18
m-Phenylene diisocyanate	4.34	0.517	8.4
2,4-Toluene diisocyanate	1.98	0.166	11.9
2,6-Toluene diisocyanate	0.884	0.143	6.11
4,4-Methylene bis (phenylisocyanate)	0.96	0.33	2.91

*Triethylamine catalyzed reactions with 1-butanol in toluene at 30.7°C.
Data Source: (Reference [23])—Burkus and Eckert, *J. Am. Chem. Soc.*, 80, 5948 (1957).

Table 4. Relatives rates of urethane formation, urea formation and cyclotrimerization reactions.

	Relative Rates		
	Urethane-Formation	Urea-Formation	Trimerization
p-Phenylene diisocyanate (PPDI)	1.85	2.15	1.5
1,5-Naphthalene diisocyanate (NDI)	0.37	1.45	—
4,4′-Methylene bis(phenylisocyanate) (MDI)	1.00	1.00	1.00
Trans-cyclohexane 1,4-diisocyanate (CHDI)	0.28	0.01	0.024
4,4-Methylene bis(cyclohexyl) diisocyanate	0.13	0.005	0.013

Data Source: (Reference [25])—Wong and Frisch, *Advances in Urethane Science and Technology*, 8, 75–92 (1981).

that the type of catalyst used has a significant influence on the reactivity of the first and the second isocyanate groups. The k_1/k_2 ratio was observed by these investigators for the reaction of PPDI with *n*-butanol catalyzed by (dibutyltin dilaurate) to be 2, as compared to 9.1 observed by Burkus and Eckert [23] using triethylamine catalyst. This means that the tin catalyst activates the second isocyanate group in PPDI nearly to the same reactivity level as the first isocyanate group.

The overall relative reaction rate data for various diisocyanates studied by Wong et al. [25] is summarized in Table 4. The first column shows the relative rate of urethane formation reaction with *n*-butanol in toluene at 50°C using dibutyltin dilaurate as catalyst; the second column presents relative rates of urea formation reaction with water in cellosolve acetate at 50°C using triethylenediamine as catalyst while the third column depicts the relative rates for the isocyanurate formation reaction or cyclotrimerization reaction, catalyzed by sym-tris(dimethylaminopropyl) hexahydrotriazine. The data for other diisocyanates is relative to the rates observed for MDI.

p-PHENYLENE DIISOCYANATE-BASED ELASTOMERS

Extensive studies on PPDI-based elastomeric polyurethanes have been carried out by several groups of investigators. Notably, groups led by K. C. Frisch [6,7,16,25] at the University of Detroit, Mercy Michigan, U.S.A. (work supported by both AKZO Chemie and E. I. du Pont de Nemours & Co., Inc.), and by C. Hepburn [9–11] at Loughborough University of Technology, U.K. have conducted highly systematic evaluations of PPDI-based polyurethanes. Later, Uniroyal Chemical Co., Inc. [12,13] who has commercialized PPDI-based prepolymers has published studies related to PPDI-based elastomeric polyurethanes and has discussed their use in many demanding applications. Carefully formulated PPDI-based polyurethane elastomers have shown a unique combination of superior dynamic properties, high resilience, good hydrolytic stability, exceptional flex fatigue, and high abrasion resistance, relative to other commercially available polyurethane elastomers. PPDI- and CHDI-based elastomers also show excellent retention of their mechanical properties at elevated temperatures and consequently are apt to perform better at elevated temperatures, compared to their TDI- and MDI-based counterparts.

PPDI can provide both thermoset and thermoplastic polyurethane elastomers with excellent mechanical and engineering properties. The rigid rod-like structure of PPDI (and CHDI) compared to the more bulky or hindered structures for MDI and TDI, is judged to be responsible for the relative superior mechanical and dynamic properties of the PPDI (and CHDI) based urethanes [8].

The thermoset elastomers are prepared through a prepolymer route. The prepolymers are prepared easily by reacting an excess of PPDI with polyols at temperatures ranging from 80 to 135 °C. When all the hydroxyl groups of the polyol are reacted, a stable mixture of isocyanate terminated polyol molecules with some free (unreacted) PPDI, called a prepolymer, is formed. The prepolymers based on PPDI are quite stable when stored under ambient conditions in tightly sealed containers (to avoid access of moisture). The prepolymers are further processed or cured as commonly practiced with the polyurethane prepolymers prepared from other diisocyanates e.g. TDI and MDI. PPDI-based prepolymers can be cured with both the diol type of curatives such as 1,4-butanediol or 1,4-cyclohexanedimethanol (CHDM), or the diamine type of curatives, namely MBCA, methylenebis(orthochloroaniline) or Ethacure 300, di(methylthio)toluenediamine. Generally the prepolymer is heated (often up to 100 °C maximum) to lower its viscosity (to facilitate mixing of the curatives) before a preweighed quantity of a curative is added. The reaction of prepolymer-curative mixtures is often accelerated by the use of common polyurethane catalysts (tertiary amines such as triethylenediamine, or the organo-metallic catalysts such as dibutyl tindilaurate). After the curative (and catalyst) is thoroughly mixed with the prepolymer, the mixture is poured into heated molds (generally 70 to 120 °C), and the filled molds are placed in an oven for curing. The curing is generally done at temperatures of 100 to 150 °C for a certain period of time at which the mold contents are firmly set up and can be demolded without deforming. The demolded articles are then postcured or annealed, if desired, to develop the desired ultimate high properties of the elastomer and thus enhance the performance of the article.

PPDI has also been used to produce thermoplastic polyurethanes [3,14,16–18,26]. Similar to the well known MDI-based polyurethane thermoplastics, both a one step process (in which the polyol, the diisocyanate and the chain extender are mixed together) or a two step process (in which a prepolymer is first made by reacting a polyol with the diisocyanate, which is then reacted with a chain extender in a second step) have been used for making the PPDI-based thermoplastics. Although many chain extenders such as 1,4-butanediol, 1,4-cyclohexanedimethanol (CHDM), 1,4-bis(beta-hydroxyethoxy)benzene (HQEE), and 1,1′isopropylidene-bis-(*p*-phenylene-oxy)-di-s-ethanol (Dianol 22) [10,11] have been used in studying the PPDI-based thermoplastics, butanediol and CHDM remain the most common ones. The prepolymer route is also considered to be the preferred route for making the thermoplastics from PPDI, mainly due to the toxicological concerns with handling the monomeric PPDI.

It was observed that post-curing or annealing at 150 °C helps in fully developing the properties of the TPU's [18]. Since the PPDI-based TPU's must be processed at temperatures higher than their inflection temperatures (temperatures at which the DSC curve starts to show a sharp melting peak), it is suggested [18] that annealing of the injection molded articles based on the PPDI TPU's has a significant effect improving their tensile strength, tear resistance, compression set, abrasion, and heat resistance characteristics.

PROCESSING OF PPDI-BASED PREPOLYMERS

In a systematic study [30], prepolymers from different diisocyanates were prepared by reacting two moles of a diisocyanate with one mole of a polyol at 80 °C, (except for those based on the MDI, in which the MDI to polyol ratio was 3/1). All resins were cured with different curatives at 95% stoichiometric equivalents of the curatives. Generally the curative and the prepolymer were mixed at 70–80 °C (except for HQEE which was heated to 110–120 °C for melting and then mixed into a prepolymer to prevent crystallization of the HQEE). A catalyst, stannous octoate (Dabco T-9), was used in some cases. The molds and curing temperatures were held at 100–130 °C for 30 minutes to complete the cure. The set times observed with different curatives are summarized in

Table 5. Set times of PPDI-based prepolymers (3.7% NCO) with different type of curatives.

	Polyol Type								
	PTMG-2000			Polycaprolactone-2000			Polybutyleneadipate-2000		
Curative	Prepolymer Temp °C	Curative Temp °C	Set Time Minutes	Prepolymer Temp °C	Curative Temp °C	Set Time Minutes	Prepolymer Temp °C	Curative Temp °C	Set Time Minutes
1,4 Butanediol Cat. 0.02%	75	75	2	85	85	3.7	80	80	2.5
Hydroquinone di(beta-hydroxyethyl)ether (HQEE) (Cat. 0.01%)	90	110	1	120	120	2.3	100	100	2.0
1,4-Cyclohexanedimethanol (CHDM) (cat. 0.02%)	80	80	3	85	85	4.3	85	85	3.0
1,2-Bis (aminophenylthio)ethane (Cyanacure)® **	85	85	2	95	95	4.5	95	95	2.0
Di-(methylthio) toluene diamine (Ethacure-300)® **	80	80	1.5	85	85	4.4	80	80	2.0

*Available from Air Products and Chemical.
**Available from Ethyl Corp.
Data Source: (Reference [30])—Private communications E. I. du Pont de Nemours Inc.

Table 5 to illustrate the relative reactivity of different curatives with PPDI-based prepolymers obtained with different types of polyols.

As evident in Table 5, both the typical diol type and the amine type curatives can be used and such curatives can provide manageable processing times with the PPDI prepolymers derived from all kinds of soft segments or polyols.

The processing behavior of the PPDI-based resins relative to those based on other types of diisocyanates is summarized in Table 6. Here data is provided only for 1,4-butanediol as the curative. Table 6 also summarizes the physical properties of the elastomers. The superior properties, especially the % elongation, tear strength, and rebound are evident for the PPDI-based elastomers. Similar observations are reported by many other investigators [6,7].

MORPHOLOGY OF PPDI-BASED ELASTOMERS

The morphology of PPDI-based elastomers has been studied [26] by using wide and small angle X-ray scattering, DSC and density measurements. In one study [26] polytetramethylene ether glycol (PTMG) was used as the flexible or soft segment while the hard segments were comprised of PPDI and 1,4-butanediol. The hard segment content was systematically varied by using PTMG of different molecular weight ranging from 650 to 2000. The DSC data are summarized in Table 7.

The three endothermic peaks designated as Peak 1, Peak 2, and Peak 3 in Table 7 provide an insight into the morphology of PPDI-based elastomers. These observations are similar to those made by others [27] on MDI and TDI based polyurethanes. Peak 1 corresponds to the melting of the flexible soft segment, PTMG. This peak observed in samples A and B is significantly lower in temperature and weaker than those reported for the pure PTMG at 43 °C [28]. The interference of hard segments in the crystallization of the soft segment is alleged to cause this endotherm to weaken in intensity and appear at a lower temperature. As the number of the repeat units decreases in samples C and D, the crystallinity of the soft segment (Peak 1) disappears.

Peaks 2 and 3 have been attributed to the hard segments. Peak 2 becomes sharper and more intense as the weight fraction of hard segments increases. This peak is attributed to somewhat poorly formed hard domains. Peak 3 is attributed to relatively well formed hard segment domains and as the hard segment content increases this peak appears at higher temperatures, perhaps due to decreased melting point depression effect of reduced soft segment acting as impurity.

Table 6. Typical processing of prepolymers based on different diisocyanates and cured with 1,4-butanediol at 95% stoichiometry and properties of their elastomers.

Isocyanate		MDI			CHDI	
Polyol	PTMG	PCL	PBA	PTMG	PCL	PBA
Prepolymer NCO %	6.15	6.09	6.13	3.31	3.61	3.55
Dabco T-9 Wt. %	0.03	0.02	0.03	0.02	0.08	—
Temp ºC, Prepolymer	85	85	80	80	85	100
Temp ºC, Curative	85	85	80	75	85	100
Set Time (Min.)	3	1	2	3	2.5	2
Molding Time Mins/ºC	30/100	30/100	30/100	30/120	30/100	30/100
Postcure, Hrs/ºC	24/100	24/100	24/100	18/85	24/100	24/100
Density g/cm³	1.02	1.12	1.11	1.06	1.01	1.02
Hardness Shore A	63	80	85	88	88	90
Hardness Shore D	20	34	35	33	42	42
10% Modulus, Psi	147	281	343	415	490	989
Tensile Strength, Psi	2498	5772	5378	3958	3215	4269
Elongation, %	401	520	447	314	770	595
Split Tear, Pli	31	199	477	284	470	436
Bashore Rebound	19	41	46	59	54	53
Compression Set (22 Hrs. @ 70ºC)	25	47	55	32	30	52

Table 6. (continued).

Isocyanate	TODI			NDI			PPDI		
Polyol	PTMG	PCL	PBA	PTMG	PCL	PBA	PTMG	PCL	PBA
Prepolymer NCO %	3.34	3.19	3.37	3.38	3.23	3.47	3.76	3.27	3.37
Dabco T-9 Wt. %	0.07	0.07	0.02	0.01	0.02	0.02	0.02	0.03	0.02
Temp ºC, Prepolymer	80	85	100	70	85	90	75	85	80
Temp ºC, Curative	75	85	100	70	85	90	75	85	80
Set Time (Min.)	3	5.0	1.5	3	3.5	1	2	3.7	2.5
Molding Time Mins/ºC	30/120	30/100	30/100	30/120	30/100	30/100	30/120	30/100	30/100
Postcure, Hrs/ºC	18/80	24/100	24/100	18/80	24/100	24/100	18/80	24/100	24/100
Density g/cm³	1.06	1.21	1.14	1.08	1.21	1.08	1.03	1.07	1.11
Hardness Shore A	80	90	68	90	89	92	85	90	96
Hardness Shore D	25	30	27	40	39	48	33	41	45
10% Modulus, Psi	275	349	346	328	665	903	347	457	831
Tensile Strength, Psi	4086	4165	5964	3647	6314	6064	4759	5607	5911
Elongation, %	540	592	431	304	595	620	626	635	800
Split Tear, Pli	154	180	261	59	242	403	94	245	649
Bashore Rebound	55	48	32	60	58	47	67	58	55
Compression Set (22 Hrs. @ 70ºC)	53	45	48	48	39	39	28	49	57

PTMG = Polytetramethylene ether glycol; PCL = Polycaprolactone glycol; PBA = Polybutyleneadipate glycol; all polyols have nominal molecular weight of 2000.
Data Source: (Reference [30])—Private communications E. I. du Pont de Nemours Inc.

Table 7. DSC peaks of PPDI-based elastomers.

Sample	M. Wt. of PTMG	% NCO	W (h)*	Pot Life** (min)	Density	Peak 1 °C	Peak 2 °C	Peak 3 °C
A	2000	3.60	0.14	10	1.044	15		200
B	1500	4.69	0.17	7	1.077	2	184	231
C	1000	6.30	0.24	4	1.096	—	186	240
D	650	8.60	0.33	2	1.134	—	190	242

*W (h) is the weight fraction of the hard segments.
**Prepolymer and 1,4-butanediol mixed at 70°C.
Data Source: (Reference [26])—Kazmierczak, Fornes, Buchanan, and Gilbert, *J. Polymer Sci., Part B: Polymer Physics*, 27, 2173–2187 (1989) and ibid 27, 2189–2202 (1989).

Samples A through D were also annealed in a sister study [26] at 125, 150, 175, and 200°C. Analysis of DSC, SAXs and WAXS data for the annealed and unannealed samples shows a profound effect of annealing PPDI elastomers. The phase separation becomes facile and the DSC peaks 2 and 3 appear at higher temperatures as the annealing temperatures are increased. Based on these observations, a more systematic study of the effects of annealing PPDI elastomers on their mechanical and dynamic properties is clearly warranted.

PROPERTIES OF PPDI ELASTOMERS

An early study by Blaich and Sampson [2] compares the mechanical properties of methylene bis(orthochloroaniline), MBCA cured elastomers from a series of diisocyanates, including PPDI. These investigators were unable to clearly show some of the attributes of the PPDI elastomers except that PPDI provided elastomers with the highest hardness at the same level of hard segment concentration. Some of the key attributes of the PPDI elastomers as studied by later investigators are discussed in the following sections.

TOUGHNESS OF PPDI-BASED ELASTOMERS

A unique combination of high tensile strength with high break elongations is observed by many investigators with the PPDI elastomers as is obvious from the data presented in Table 6. Similar observations were made by Singh et. al. [32] in a study comparing the properties of the PPDI, MDI, and TDI elastomers as shown in Table 8. All elastomers

Table 8. Relative properties of PPDI, TDI, and MDI Elastomers[a] (90 Shore A).

Prepolymer Type	PPDI[b]		MDI	TDI
Curatives	1,4BD	CHDM[c]	1,4BD	MBCA[d]
Properties				
Shore A Hardness	93	88	90	90
Modulus, 100% MPa	7.10	6.07	7.45	7.93
300% MPa	10.62	9.66	16.97	17.03
Tensile, MPa	35.31	36.0	33.03	38.21
% Elongation	633	570	410	370
Compression Set, %, 70°C	22	20	30	28
Method B 100°C	44	42	63	68
Trouser Tear (D-1938)	—	—	—	—
KN/m 70°C	23.65	25.92	21.89	21.54
100°C	10.16(43)[e]	11.38(44)	4.73(22)	6.30(29)
% Bashore Rebound, %	64	66	60	45
Tabor Abrasion[f]	142	20	157	175

[a]All elastomers are based on the same polyether polyol soft segment.
[b]PPDI prepolymer is 3.5% NCO content.
[c]CHDM: 1,4-Cyclohexanedimethanol cured.
[d]MBCA: 4,4-Methylene-bis-(2-chloraniline) cured.
[e]% of 70°C Trouser Tear.
[f]H-18, 1000 g load, 5000 cycles, wt. loss in mg.
Data Source: (Reference [32])—Singh, Rosenberg, Chin, and Gajewski, *Proc. Utech Asia'95* Singapore, May, 1995.

are based on the same soft segment, PTMG and are compared at the same elastomer hardness. W. C. Whelchel [15] compared the product of the tensile strength and the ultimate elongation as a rough measurement of the break energy or the toughness of elastomers based on different diisocyanates. These data are presented in Figure 1. Although the toughness is generally defined as the area under the stress-strain curve, the data shown in this figure is a simpler way to show the superior toughness of PPDI elastomers.

HIGH TEMPERATURE PROPERTIES OF PPDI ELASTOMERS

Barikani and Hepburn [11] studied elastomers from different diisocyanates based on polycaprolactone soft segment and cured with butanediol, using both the DMTA and DSC measurements. From their studies they concluded that the short-term thermal stability of elastomers produced from different diisocyanates is in the order depicted below:

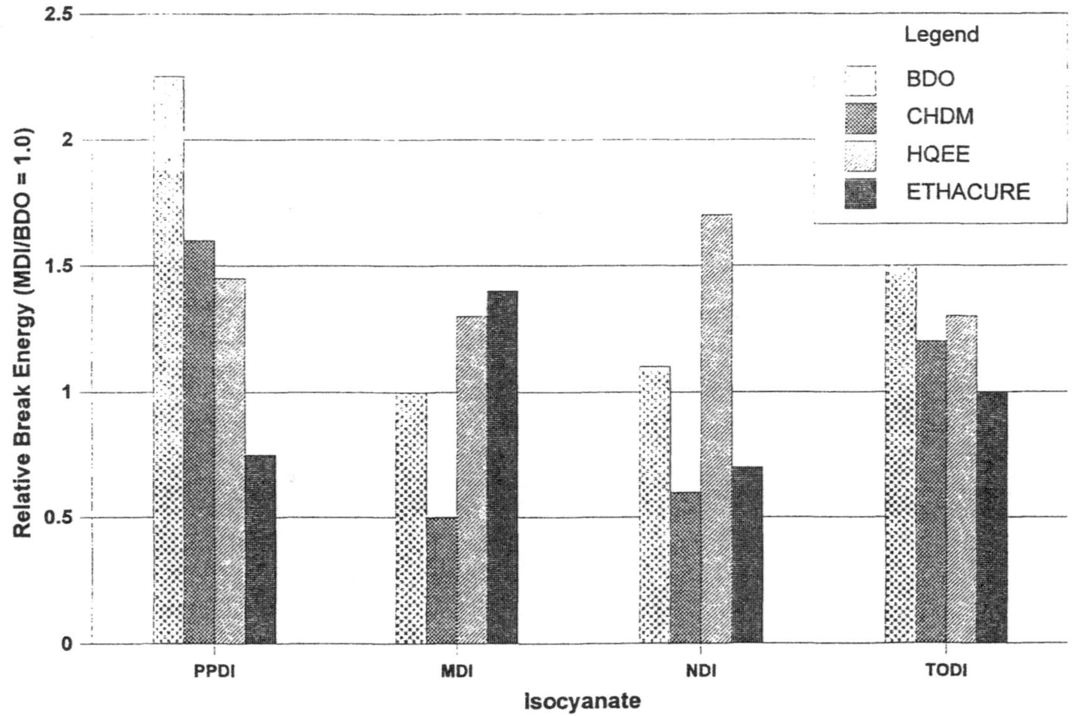

Figure 1. Relative break energy of different elastomers. Data Source: (Reference [15])—Moore, Whelchel, and Russell, Proceedings of the SPI, Polyurethanes World Congress, *Nice, France, September 1991.*

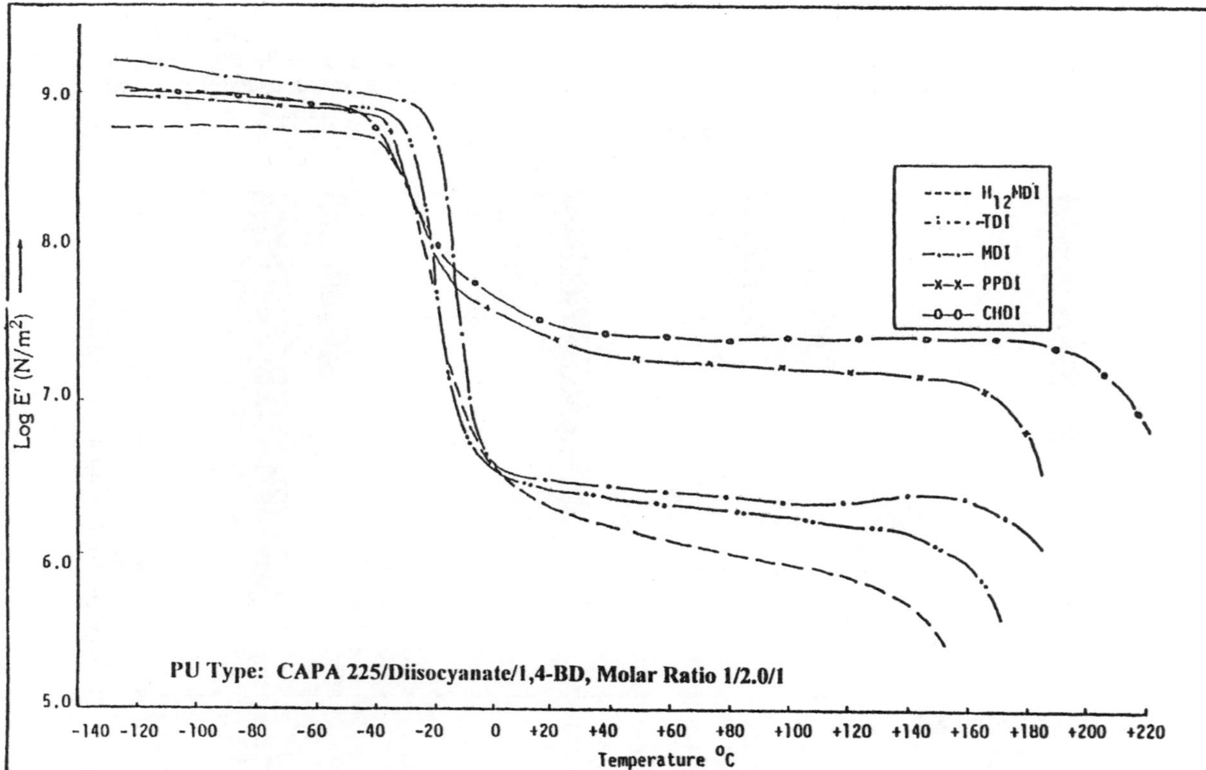

Figure 2. *Storage modulus as a means of observing the dependence of temperature stability on diisocyanate structure (PU type: Capa 225/diisocyanate/1,4-BD: Molar ratio: 1/2.6/1). Data Source: (Reference [11])—Barikani and Hepburn, Cell. Polym., 6, (3), 41–54 (1987).*

Table 9. Percent retention of properties at elevated temperatures for 1,4-BD cured elastomers.

Diisocyante	MIDI			TODI			NDI			PPDI		
Polyol	PTMG	PCL	PBA	PTMG	PCL	PBA	PTMG	PCL	PBA	PTMG	PCL	PBA
10% Modulus												
25ºC (Psi)	113	81	400	162	114	188	209	219	557	153	160	506
100ºC (Psi/% Ret.)	74/65	34/42	93/23	128/79	80/70	89/47	167/80	194/89	226/47	98/64	78/49	163/32
150ºC (Psi/% Ret.)	62/55	31/38	29/7	116/72	56/49	84/45	101/48	118/54	218/39	83/54	101/63	139/27
Tensile Strength												
25ºC (Psi)	4212	4303	3534	3370	3950	5293	4863	5428	5126	4998	5193	5168
100ºC (Psi/% Ret.)	1650/39	1823/42	676/19	1391/41	775/20	2801/34	1456/30	2613/49	2805/55	2063/41	2180/42	2846/55
150ºC (Psi/% Ret.)	766/18	458/11	211/6	879/26	221/6	845/16	915/19	848/16	1485/29	969/19	1225/24	931/18
Elongtion %												
25ºC (Psi)	740	781	1001	1222	970	1032	946	979	881	1091	1097	1043
100ºC (Psi/% Ret.)	812/110	1214/181	679/68	1074/88	742/76	1011/98	632/67	1007/103	1000/114	923/85	1028/94	1400/34
150ºC (Psi/% Ret.)	773/64	955/122	114/11	816/67	799/82	1262/122	529/56	757/77	870/99	662/61	976/89	1038/100

Data Source: (Reference [30])—Private communications E. I. du Pont de Nemours Inc.

CHDI > PPDI > MDI > TDI > HMDI

Their conclusion is evident from Figure 2, where the storage modulus of different elastomers is shown as a function of temperature. Similar observations were made by Syed et al. [8] when they studied the storage shear modulus as a function of temperature for elastomers based on different diisocyanates, using a rheometrics mechanical analyzer.

The effects of short term exposure to elevated temperatures on the mechanical properties, such as the tensile strength, 10% modulus and elongation of PPDI elastomers relative to other commercially important systems are listed in Table 9 [30]. The retention of these properties by the PPDI elastomers is similar to those of the TODI and NDI based systems and superior to those of the MDI counterparts.

A comparison of PPDI and MDI-based TPU's [3] shows that under 150°C air aging, PPDI elastomers depict a superior heat resistance than their MDI counterparts. The hardness drop at elevated temperatures observed for the PPDI based TPU's is less than that for the MDI

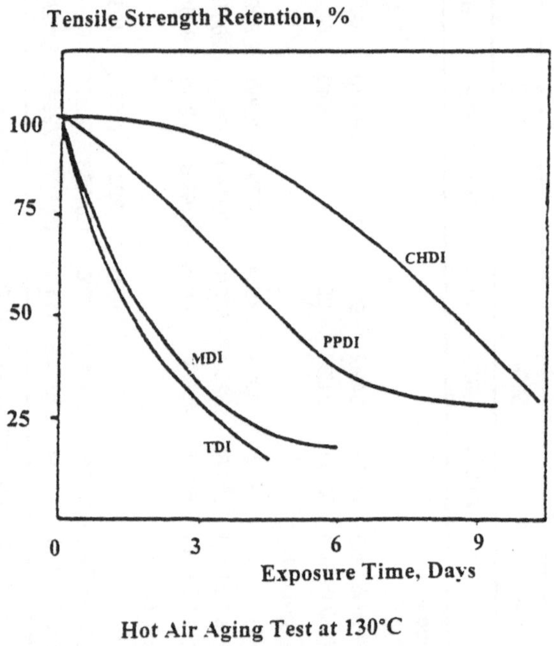

Figure 3. Tensile strength retention of various ether based polyurethane elastomers. Data Source: (Reference [8])—Syed, Muys, and Kuijpers, Proceedings UTECH'86, *The Hague, March 1986, page 88.*

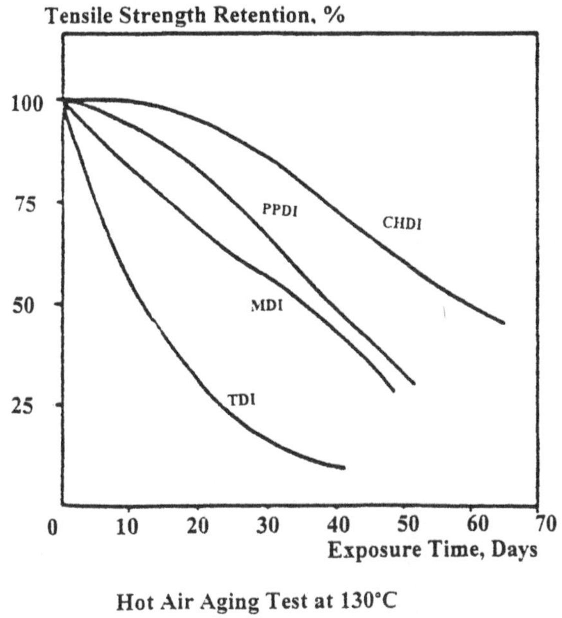

Figure 4. Tensile strength retention of various ester-based polyurethane elastomers. Data Source: (Reference [8])—Syed, Muys, and Kuijpers, Proceedings UTECH'86, *The Hague, March 1986, page 88.*

based TPU's. Similar superior retention of the hardness, tensile strength, and tear properties after aging at elevated temperatures by the PPDI based elastomers is documented by other investigators [9,10]. Figures 3 and 4, which show the level of tensile strength retention for polyester and polyether elastomers based on different diisocyanates [8] show that both the CHDI and PPDI elastomers retain significantly higher tensile strength on heat aging.

Similar observations are made in other studies [6,7] in which the modulus values at 300% elongation for elastomers prepared from different diisocyanates are compared. As Figures 5 and 6 show, PPDI, CHDI and NDI elastomers retain higher modulus values at elevated temperatures, compared to the MDI elastomer in both the polycaprolactone and polytetramethylene ether glycol-based elastomers.

The compression set values shown in Table 8 especially those observed at 100°C are significantly lower than those for the TDI and MDI counterparts of similar hardness and provide further evidence of a superior high temperature performance capability expected from the PPDI elastomers.

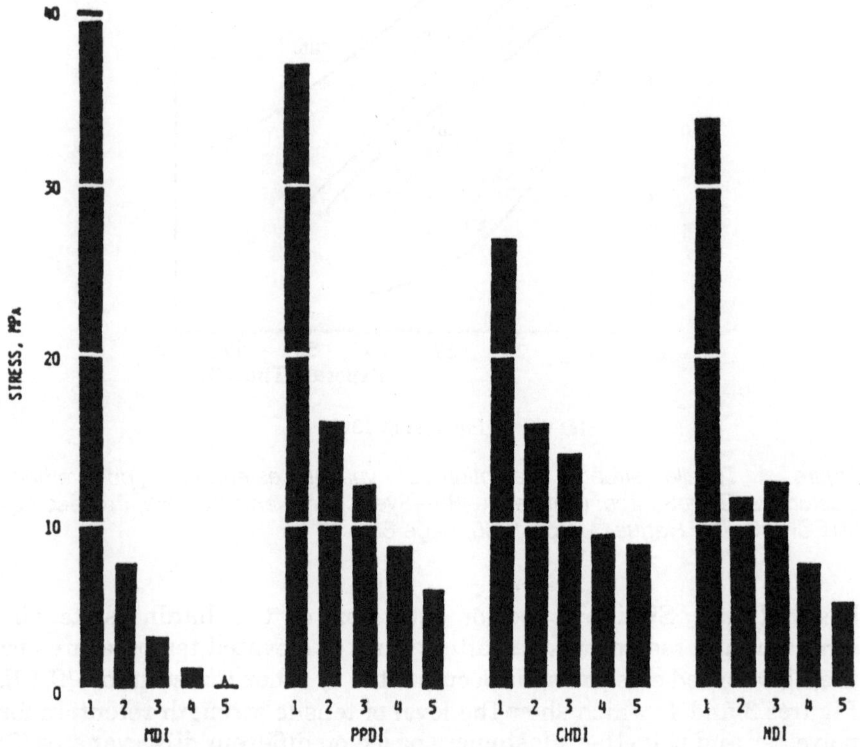

Temp.: 1 = 20; 2 - 25, 3 = 70; 4 = 121; 5 = 150°C

^SAMPLE MELTED

Figure 5. Stress properties at 300% elongation of polycaprolactone polyol elastomers at various temperatures. Data Source: (Reference [6])—Wong and Frisch, Advances in Urethane Science and Technology, *8, 75–92 (1981).*

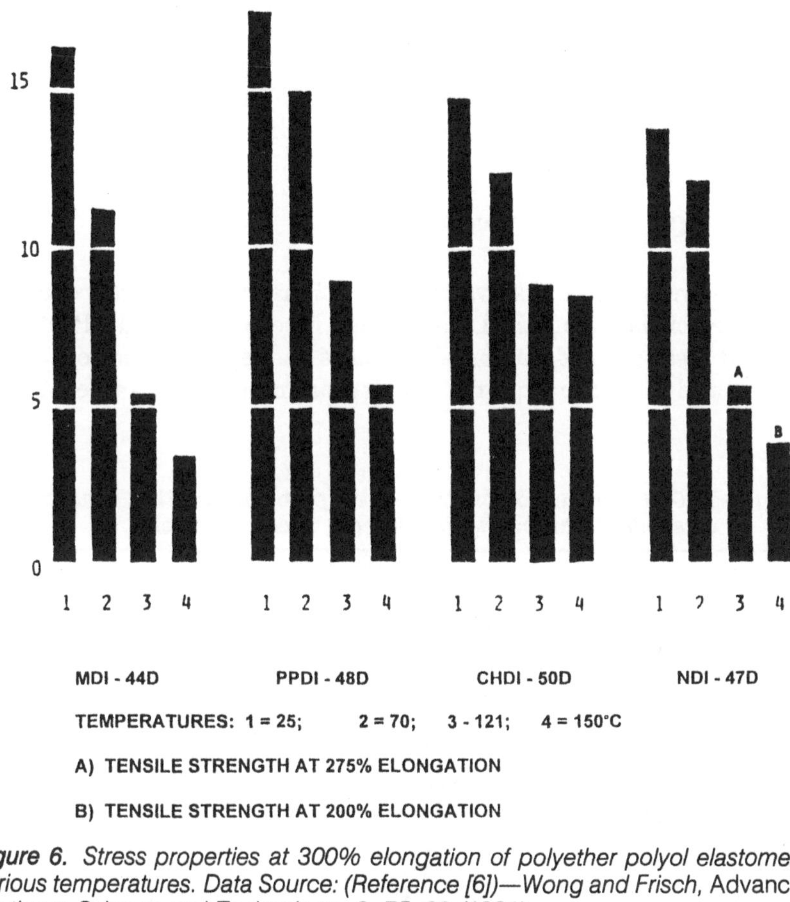

Figure 6. *Stress properties at 300% elongation of polyether polyol elastomers at various temperatures. Data Source: (Reference [6])—Wong and Frisch,* Advances in Urethane Science and Technology, *8, 75–92 (1981).*

HYDROLYTIC STABILITY

The hydrolytic stability of PPDI elastomers has been studied by several investigators [3,8,9,12]. The relative hydrolytic stability of PPDI, TDI and MDI elastomers is depicted in Table 10. Elastomer samples were boiled in water for 8 days and 30 days after which they were dried and tested for their stress-strain properties [32]. The PPDI elastomers cured with diols have shown good hydrolytic resistance, which is similar to MDI/diol elastomers, while the TDI/diamine elastomers did relatively less well under the same test conditions.

FLEX FATIGUE RESISTANCE

A limited number of observations [15,32] have been made regarding flex life performance of PPDI elastomers. Moore et al. at du Pont have compared Ross Flex data for elastomers made from different diisocyanates and have reported the results summarized in Table 11. The

Table 10. A comparison of stress-strain properties of PTMG-based elastomers after boiling in water (90 Shore A elastomers).

Prepolymer Type	PPDI		TDI	MDI
Curatives	1,4-Butanediol	CHDM	MBCA	1,4-Butanediol
Ungaged Properties (MPa)				
100% Modulus	7.10	6.07	7.45	7.93
300% Modulus	10.62	9.66	16.97	17.03
Tensile Strength	35.31	36.07	33.03	38.21
% Elongation	633	570	410	370
8 Days in Boiling Water (MPa)				
100% Modulus	5.45	4.87	3.86	4.97
300% Modulus	8.62	7.86	3.86	9.79
Tensile Strength	20.97	24.96	4.00	29.10
% Elongation	790	730	270	580
30 Days in Boiling Water (MPa)				
100% Modulus	5.31	4.69	—	4.45
300% Modulus	8.21	7.45	—	8.00
Tensile Strength	17.03	22.41	2.75	17.72
% Elongation	780	830	50	630

All samples were tested at room temperature after drying.
Data Source: (Reference [15])—Moore, Whelchel, and Russell, *Proceedings of the SPI, Polyurethanes World Congress*, Nice, France, Sept. 1991.

Table 11. Flex fatigue data using Ross flexometer for elastomers based on different diisocyanates.

Formulation	Cycles to Failure
PTMG/PPDI/BDO	>1,000,000
PTMG/PPDI/HQEE	>950,000
PTMG/PPDI/CHDM	30,000
PTMG/TODI/BDO	60,000
PTMG/NDI/BDO	60,000
PTMG/MDI/BDO	12,000

Data Source: (Reference [15])—Moore, Whelchel, and Russell, *Proceedings of the SPI, Polyurethanes World Congress*, Nice, France, Sept. 1991.

Table 12. Flex fatigue (TEXUS FLEX) data cycles to failure.

Prepolymer	Curative	Shore A	Texus Flex Cycles at 30 Deg.	Texus Flex Cycles at 45 Deg.
PPDI/Polyester	BDO	94	6480+	4222+
"	A-240*	88	6480+	4222
"	HQEE/A-240*	90	6480	2125
PPDI/PTMG	BDO	96	6480	450
"	A-240*	94	1500	90
"	HQEE/A-240*	95	1500	102
TDI/PTMG	MBCA	95	10	0.5
"	MBCA	90	18	6
MDI/PTMG	BDO	95	12	1
"	BDO	90	20	8

*Special grade of 1,4-cyclohexanedimethanol (CHDM).
All samples had an initial 2-mm cut at a 10 degree angle. Measurements made at 70°C.
Data Source: (Reference [32])—Singh, Rosenberg, Chin, Gajewski, *Proceedings Utech Asia'95* Singapore, May, 1995.

cycles to failure for the PPDI elastomers stand out significantly over the results obtained with other elastomers. Similarly several elastomers were tested by Singh et al. of Uniroyal Chemical Company [32] using the Texus Flex instrument. Elastomers based on different diisocyanates were tested at 30 degrees and 45 degrees angle of inclination (higher angle applies higher strain energy to samples under test). The data in Table 12 again shows a significant superiority of the PPDI elastomers over their MDI and TDI counterparts of similar hardness.

DYNAMIC PERFORMANCE

Dynamic mechanical performance of PPDI elastomers has been a subject of several studies [3,6,10,11,15]. In general the CHDI and PPDI based elastomers have a wider plateau region when their dynamic storage modulus is measured as a function of temperature. Also the low temperature loss tangent (tan ∂) peaks of PPDI elastomers have lower values, and shift to lower temperatures, compared to the TDI or MDI elastomers of similar hardness. Barikani and Hepburn [11] have measured the tangent ∂ peak values at, what they call, the transition temperature (T_g), the storage modulus (log E'), and T°C, the maximum temperature at which the storage modulus value will remain constant. These data are summarized in Table 13 for elastomers prepared from different diisocyanates. The lower values for tan ∂, and the T_g coupled with higher values for log E' and T°C all indicate the superior dynamics observed for the CHDI and PPDI-based elastomers relative to the other diisocyanates studied.

The simple measurements for Bashore rebound as shown in Tables 6 and 8 also indicate the superior dynamic properties of PPDI based cast and thermoplastic elastomers, as is shown graphically in Figure 7 by

Table 13. Dynamic properties of different elastomers.

Diisocyanate Type Used	Tan ∂	T_g°C	Log E' (N/m²)	T°C
CHDI	0.285	−33	7.4	181
PPDI	0.375	−30	7.1	155
MDI	1.1	−15	6.4	150
TDI	1.0	−21	6.1	135
HMDI*	0.575	−22	6.2	100

*Hydrogenated MDI.
Data Source: (Reference [11])—Barikani and Hepburn, *Cell. Polym.*, 6, (3) 41–54 (1987).

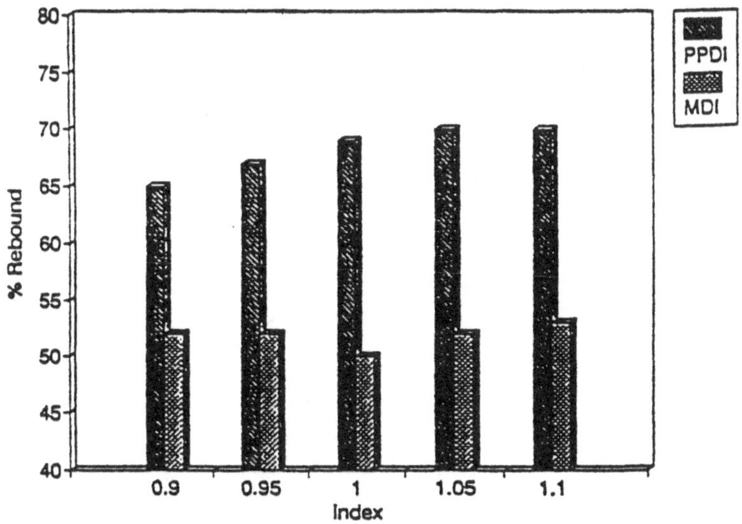

Figure 7. Resilience of PPDI/PTMEG compared to MDI/PTMEG systems. Data Source: (Reference [14])—Moore, Russell, and Whelchel, Rubber and Plastic News, *21 (5), 15–18, (1991).*

Moore et al. of du Pont where percent rebound of PPDI and MDI thermoplastic elastomers is depicted as a function of their isocyanate index.

The dynamic performance of PPDI elastomers has been studied by several investigators [8,11,12,29]. Using rheometrics viscoelastic test equipment, Singh [12] and coworkers calculated the power loss values for elastomers of equal hardness based on a variety of diisocyanates. Using an equation:

$$\text{Power loss, } E^{\circ} = W/2 \cdot Y^{\circ 2} \cdot G''$$

where W = frequency in radians per second, Y° = maximum strain, and G'' = dynamic loss modulus.

The lower the power loss value and more constant that value remains over a wide temperature range, the better an elastomer is in its ability to resist the hysteresic heat buildup under dynamic load application.

Figure 8 shows power loss curves for PTMG based elastomers of similar hardness (90 Shore A). The TDI elastomer was cured with MBCA while the MDI and PPDI prepolymers were cured with 1,4-butanediol. The data shows relatively very low power loss values, which stay stable and flat over a wider temperature range for the PPDI elastomers, in-

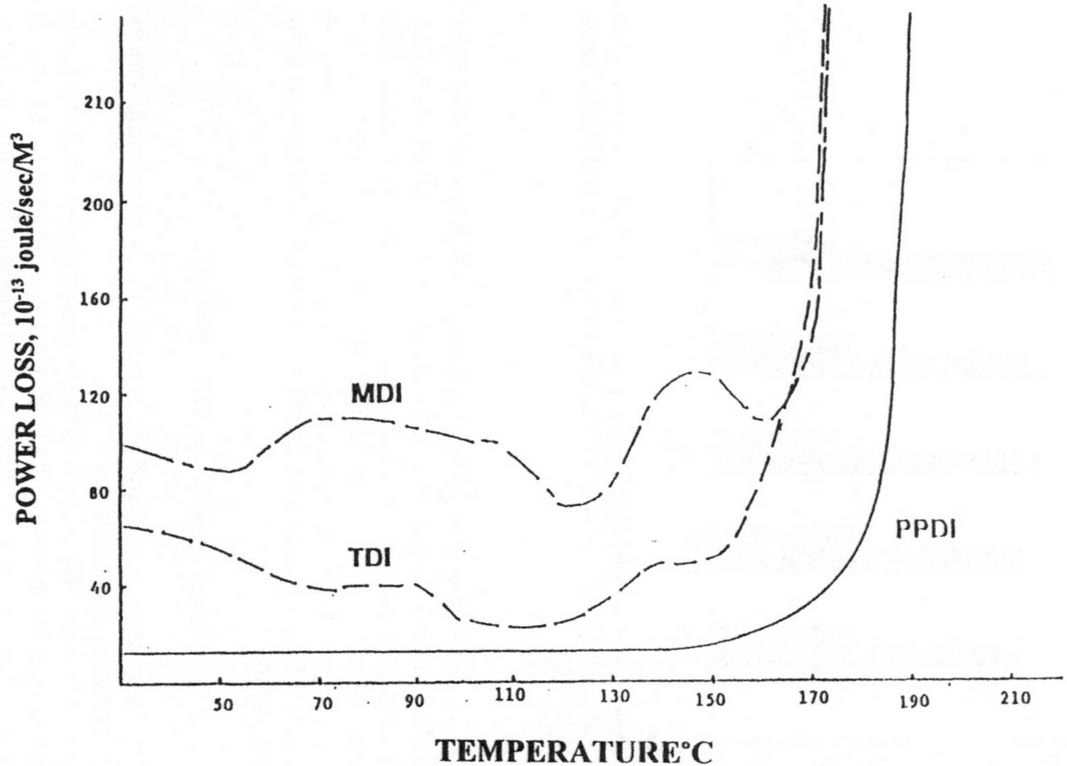

Figure 8. Power loss of 90 Shore A Elastomers, Comparison on Rheometric Test. All elastomers based on the same type of polyether soft segment. Conditions: frequency, 62.8 rad. per sec.; stress, 0.01 MPa; strain, 1%. Data Source: (Reference [12])—Singh, Palinkas, Skirkanich, Proceedings ACS Rubber Div., *Nov. 1991.*

dicating significantly lower hysteretic heat buildup expected from the PPDI elastomer under dynamic loading conditions.

APPLICATIONS

PPDI elastomers have been tried in a variety of commercial applications to assess their cost effectiveness relative to other polyurethane elastomers. Although such applications are expected to continue to grow, examples of such applications in which the PPDI elastomers have shown an excellent fit are summarized below [32]. Most of these applications demand a combination of performance attributes such as high temperature resistance, high cut and tear resistance, hydrolytic stability, excellent dynamic properties, and high abrasion resistance for which the PPDI elastomers have shown a superiority.

Application	Performance Requirements
Seals and gaskets	Low compression set at elevated temperatures, and cut and tear resistance.
Hydrocyclones	Abrasion resistance, and hydrolytic stability
Rolls	Low hysteresis, and cut and tear resistance
Tires and wheels	Low hysteresis, fatigue resistance, and cut and tear resistance
Pump parts (diaphragms, impellers)	Abrasion resistance and fatigue resistance

REFERENCES

1. Pyman, F. L. 1913. *J. Chem. Soc.,* 103:852.
2. Blaich, C. F. Jr. and A. J. Sampson. 1960. *Proceedings A.C.S. Div. of Rubber Chem.,* September, New York, NY.
3. Maeda, Y., Y. Ohbuchi and S. Sato. 1989. *Proceedings of the SPI 32nd Annual Technical/Marketing Conference,* October.
4. Schollenberger, C. S. Ger. Patents 1045,643 (1958) and 1049,090 (1959), and Brit. Pat. 901,910 (1962).
5. Murphy, W. T. 1962. U.S. Pat. 3,3033,825.
6. Wong, S. W. and K. C. Frisch. 1981. *Advances in Urethane Science and Technology,* 8:75–92.

7. Wong, S. W., W. J. Juang, V. Olgac, K. C. Frisch, G. Heinrichs and P. Hentschel. 1981. *A.C.S. Symposium Series,* 172:419–431.
8. Syed, E., J. Muys and K. G. Y. A. S. Kuijpers. 1986. *Proceedings UTECH'86 Polyurethane Industry's International Conference,* The Hague, March:86–89.
9. Hepburn, C. 1992. *European Rubber J.,* 164(1):60–62; Barikani, M. and C. Hepburn, *Iran. J. Polym. Sci. and Tech.,* 1(1):1–5.
10. Hepburn, C. 1987. *Cell. Polym.,* 6(5):51–66.
11. Barikani, M. and C. Hepburn. 1987. *Cell. Polym.,* 6(3):41–54.
12. Singh, A., R. Palinkas and N. J. Skirkanich. 1992. *Proceedings A.C.S. Rubber Div.,* November.
13. Singh, A., A. M. Clock, J. Chin and N. J. Skirkanich. 1994. *Proceedings UTECH'94,* The Hague, March.
14. Moore, R., D. Russell and W. C. Whelchel. 1991. *Proceedings A.C.S. Rubber Div.,* May; Moore, R., D. Russell and W. C. Whelchel, 1991. *Rubber and Plast. News,* 21(5):15–18.
15. Moore, R., W. C. Whelchel and D. Russell. 1991. *Proceedings of the SPI, Polyurethanes World Congress,* Nice, France, September.
16. Xiao, H. X., S. Yang, J. E. Kresta and K. C. Frisch. 1994. *J. Elastomers and Plastics,* 26:237–251.
17. Russell, D., R. Moore and W. C. Whelchel. 1991. *Rubber and Plastics News,* 21(5):15–18.
18. Castro, A. J., T. Plummer, P. Hentschel and W. Brodowski. 1985. *Proceedings of the SPI Annual Technical/Marketing Conf.,* pp. 305–308.
19. Whelchel, W. C. 1991. *Rubber World,* 205(6):22–25; Air Products and Chemicals News Release. 1992. *Rubber World,* 205(6):48; 1992. *J. Elastomers and Plast.,* 24(1):12.
20. Armak Co., Chicago, Illinois, News Releases. 1981. *Elastomerics,* 113(4)82–83; 1983. 115(11):11.
21. Baker, J. W. and J. B. Holdsworth. 1947. *J. Chem. Soc.,* 713; 1949. Ibid, 18:27.
22. Bailey, M. E., V. Kirss and R. G. Spaunburgh. 1956. *Industrial and Eng. Chem.,* 48:794–797.
23. Burkus, J. and C. F. Eckert. 1957. *J. Am. Chem. Soc.,* 80:5948.
24. Dyer, E., H. Taylor, S. Mason and J. Samson. 1949. *J. Am. Chem. Soc.,* 71:406.
25. Wong, S. W., W. J. Juang, V. Olgac and K. C. Frisch. 1980. *Org. Coatings and Plast. Chem.,* 43:926-931; Wong, S. W. and K. C. Frisch. 1981. *Advances in Urethanes Sci. and Tech.,* 8:75–92; Wong, S. W., W. J. Juang, V. Olgac, K. C. Frisch, G. Heinrichs and P. Hentschel. 1981. *A.C.S. Symposium Series,* 172:419–431.
26. Kazmierczak, M. E., R. E. Fornes, D. R. Buchanan and R. D. Gilbert. 1989. *J. Polym. Sci., Part B, Physics,* 27:2173–2187; Ibid. 1989. 27:2189–2202.
27. Dieter, J. W. and C. A. Byrne. 1987. *Polym. Eng. and Sci.,* 27(9):673.
28. Koberstein, J. T. and R. S. Stein. 1983. *J. Polym. Sci., Polymer Physics Ed.,* 21:1439; Koberstein, J. T. and T. P. Russell. 1986. *Macromolecules,* 19:714.
29. Imada, K., et al. 1965. *Macromol. Chem.,* 873:113.

30. Private Communications, E. I. du Pont de Nemours, Inc.
31. News Release. 1994. *Urethane Technology,* 11(2):12.
32. Singh, A., R. O. Rosenberg, J. Chin and V. J. Gajewski. 1995. *Proceedings UTECH Asia'95,* Singapore, May.
33. Demarest, C. and R. Moore. 1987. *Proceedings of the SPI/FSK, Polyurethane World Congress,* Aachen, Germany.
34. Zengel, H. G. 1980. *Proceedings of the SPI/FSK, International Conference on Cell. and Non Cell. Urethanes,* Strasbourgh (France), June.

Urethane Elastomers/Liquid Crystalline Polyurethane Blends

J. GÄHDE, M.-M. POHL, R. MIX AND R. DANY
Institute of Applied Chemistry Berlin-Adlershof
Rudower Chaussee 5
D-12484 Berlin, Germany

G. HINRICHSEN
Technical University of Berlin, Institute of Nonmetallic Materials
Englische Str. 20
D-10587 Berlin, Germany

ABSTRACT

Nonsegmented and segmented liquid crystalline polyurethanes have been studied in mixtures with conventional elastomeric polyurethanes. The nonsegmented liquid crystalline polyurethane LCPU is composed of 2,4-tolylene diisocyanate and 4,4′-bis(6-hydroxyhexoxy)biphenyl (BHHBP) as the mesogen. By incorporation of poly(butylene adipate) soft segments of molecular weight 2000 in the main chain by a prepolymer process the segmented liquid crystalline polyurethane SLCPU was obtained. Both liquid crystalline polyurethanes were mixed with conventional elastomeric polyurethanes based on tolylene diisocyanate (TDI) or 4,4′-diphenylmethane diisocyanate (MDI), poly(butylene adipates) of molecular weight 2000 or 600 and 1,4-butanediol as chain extender.

The components and blends were characterized by differential scanning calorimetry (DSC), electron microscopy and measurements of mechanical properties.

In each blend the existence of a separated phase due to TDI-BHHBP structures can be detected by DSC. LCPU is immiscible with the TDI polyurethanes but partially miscible with the MDI polyurethane containing a short chain polyester of M_n = 600 and a relatively high content of MDI-1,4 butanediol hard segments.

SLCPU is partially miscible with TDI polyurethanes without changing the mesophase formation. In blends with MDI polyurethanes the formation of mesogenic hard segment domains is reduced. The LC properties can be obtained in miscible and immiscible blends and the high tensile strength of SLCPU can be transmitted to the blends in case of partial miscibility.

INTRODUCTION

THE MOTIVATION FOR BLENDING POLYMERS COMES FROM THE NEED TO modify properties, processability and cost [1,2,3]. The new materials exhibit new and advanced properties which cannot be obtained with nonblended polymers. Compatibility in polymer blends is controlled by thermodynamics, kinetics and other external factors such as viscosity. Miscibility can be enhanced by blending homopolymers that contain groups with favourable specific interactions, by using copolymers made up of units exhibiting mutual repulsion, by optimizing sequence distribution of copolymers and by introduction of block copolymers as compatibilizers consisting of units of each of the immiscible homopolymers [4,5,6].

The application of liquid crystalline polymers as materials with advanced properties is limited because of the high development and production costs. Blending of liquid crystalline polymers with conventional polymers can be a more economical way for getting a material with improved properties. The condition for the improvement of properties is the conservation of the better properties of the liquid crystalline polymers in the blends.

Thermoplastic elastomers are linear polyurethanes. Their excellent properties are the result of the phase separation of their hard and soft segments [7–11]. The hard segments–for example based on diphenylmethane-4,4′-diisocyanate and a short chain diol–develop a strong intermolecular interaction resulting in the formation of domains with a highly ordered or even crystalline structure. The process and state of ordering will be enhanced by incorporation of mesogenic units in the hard segments which leads to a strengthened elastomeric material with good processability which is necessary for the production of high strength elastic fibers [12].

Liquid crystalline polyurethanes represent a relatively new class of polymeric materials. Tanaka et al. [13] summarized the approaches for obtaining two possible types of liquid crystalline (LC) polyurethanes, the first one with the mesomorphic unit in the main chain and the second one with this unit in the side chain. Our work is concentrated on the main chain type of LC polyurethanes.

The two most used synthetic routes are the following:

1. Reaction of mesogenic diisocyanates with commercially available short chain diols. LC polyurethanes obtained in this way have been described by Iimura et al. [14], Mormann et al. [15–21] and Kossmehl et al. [22].

2. Reaction of mesogenic diols with commercially available diisocyanates. This way has been followed by Tanaka et al. [23–25], Lorenz et al. [26] and MacKnight and coworkers [27–29]. Sato et al. [30] used diisocyanates blocked with phenol while Kricheldorf et al. [31,32] applied dichloroformates with mesogenic diamines.

All the liquid crystalline polyurethanes mentioned above consist of hard segments only and have no elastomeric properties. The materials are characterized by a high modulus and a low ductility. Blending these liquid crystalline polyurethanes with other polyurethanes or polymers may be a way of getting a more elastic material for future technological applications.

Another way for increasing the elasticity of the LC polyurethanes of the hard segment type is the incorporation of soft segments in the main chain. Lorenz et al. [26], Penczek et al. [33–35], Tang et al. [36] and He et al. [37] have described liquid crystalline polyurethanes based on poly(tetramethylene oxides) and chain extended with a biphenyl type mesogenic diol. In our group we have synthesized segmented liquid crystalline polyurethanes based on the same biphenyl type mesogenic diol in the hard segment, but with poly(butylene adipates) as soft segments [38].

The knowledge of structure and properties of blends of liquid crystalline polyurethanes with conventional polymers is very limited up to now. Only investigations by Kressler and Kammer [39] are published, which describe blends of a liquid crystalline polyurethane, synthesized by MacKnight and his coworkers, with SAN polymers. The liquid crystalline polyurethane is composed of 2,4-tolylene diisocyanate and the mesogen 4,4′-bis(6-hydroxyhexoxy) biphenyl. Its glass transition temperature remains unchanged in the mixtures up to a SAN content of 70 wt-%. The two polymers exhibit phase separation, but the interaction between the acrylonitrile and urethane groups leads to a more increased dispersity of the components within the blend resulting in better mechanical properties.

We employed nonsegmented and segmented LC polyurethanes containing the same mesogenic hard segments composed of 4,4′-bis(6-hydroxyhexoxy)biphenyl and 2,4-tolylene diisocyanate. This study deals with the blending of these LC polyurethanes with conventional elastomeric polyurethanes. The aims of our work are the conservation of liquid crystalline properties in the blends and research on the influence of intermolecular interaction and partial compatibility.

EXPERIMENTAL

Materials

Dioxane (BASF) and *N,N*-dimethylformamide (DMF, Merck) were purified by refluxing 5 h in the presence of 4,4′-diphenylmethane diisocyanate followed by distillation or vacuum distillation respectively. *n*-Butanol was used as a reagent without purification. All solvents were stored over molecular sieves.

2,4-Tolylene diisocyanate (TDI, Merck) was used as received.

4,4′-Diphenylmethane diisocyanate (MDI, BASF Schwarzheide) and 1,4-butanediol (Merck) were purified by vacuum distillation and 1,4-butanediol was stored over molecular sieves.

Poly(butylene adipates) (PBA, BASF Schwarzheide, M_n = 600, 1000, 2000) were dried before use in vacuo for 1 h at 160 °C in a rotary evaporator. The water content was found to be < 0.05% by Karl-Fischer titration.

The synthesis of the mesogenic chain extender 4,4′-bis(6-hydroxyhexoxy)biphenyl (BHHBP) was carried out in a manner similar to the one reported by Stenhouse et al. [29]. 4,4′-Dihydroxybiphenyl (Merck) and 6-chloro-1-hexanol (Merck) have been used as received.

Synthesis:

A total of 1 l *n*-butanol, 93.2 g (0.5 mol) of 4,4′-dihydroxybiphenyl and 40 g NaOH (1 mol) were placed in a 2 l four-necked flask equipped with a dropping funnel, reflux condenser, thermometer and mechanical stirrer. The mixture was heated to boiling temperature and then 137 g (1 mol) of 6-chloro-1-hexanol were added dropwise. The reaction mixture was refluxed for 16 h. After cooling, the precipitate was filtered and washed with distilled water and recrystallized from 1-butanol (yield 80%, m.p. 172–173 °C). The purity of the recrystallized BHHBP was confirmed by HPLC to be > 99%. Before use the BHHBP was dried at 60 °C for 1 h in a vacuum chamber.

Poly(tetrahydrofurane) (PTHF, BASF, M_n = 600) was dried in vacuo for 5 h at 70 °C in a rotary evaporator. After this treatment the water content was < 0.05%.

Synthesis of Polyurethanes

Synthesis of the Nonsegmented Liquid Crystalline Polyurethane LCPU

LCPU was synthesized in solution by the reaction of BHHBP with

2,4-tolylene diisocyanate. The synthesis has been described in detail by Stenhouse et al. [29]. The solution was cooled to 50 °C and poured into methanol under vigorous stirring. The white precipitated polymer was filtered, purified by Soxhlet extraction with acetone for 5 h and dried in vacuo.

Synthesis of the Segmented Liquid Crystalline Polyurethane SLCPU

Conventional polyurethane prepolymer synthesis conditions [10] were used for the preparation of the segmented liquid crystalline polyurethanes. 105.5 g (0.05 mol) of dried poly(butylene adipate) M_n 2000 (BA20) were added to a 1000 ml four-necked flask equipped with a mechanical stirrer, thermometer, gas inlet and condenser with drying tube. The polyester was heated under nitrogen to 60 °C and 43.33 g (0.25 mol) of 2,4-tolylene diisocyanate (T) were added quickly. The reaction mixture was stirred for 30 min at 60 °C. Then 200 ml dioxane and 77.32 g (0.2 mol) of BHHBP as solid were added. The mixture was stirred under nitrogen for 20 h at 80 °C and for 1 h at 100 °C. DMF was added to keep a clear solution with a viscosity low enough for stirring. The obtained solution had a polymer content of 20–25%.

Synthesis of the Conventional Polyurethanes

Polyurethanes without mesogenic units were synthesized by the conventional prepolymer process described by Eisenbach et al. [10].

The composition of the synthesized polyurethanes as well as the results of elemental analysis and SEC are listed in Table 1. The results of nmr investigations are discussed in Reference [38].

Preparation of Polyurethane Blends and Films

The obtained polymer solutions with a polymer content of 20–25 wt-% were diluted with DMF to a polymer content of 10–15%. The composition of the resulting solvent was 60 wt-% DMF and 40 wt-% dioxane. Polymer solutions of liquid crystalline and conventional polymers were mixed to obtain the polymer blend ratio.

The blend solutions were cast on glass plates, heated to about 80 °C, stored 16 h at room temperature in an extractor hood and dryed 24 h at room temperature and 1 h at 100 °C in vacuo. The films, removed from the glass under water, had a thickness in the range of 70–100 μm.

Table 1. Investigated polyurethanes, composition and molecular weights.

	Hard Segment Content	Molecular Weight (SEC) Solvent: DMF; Temperature: 25°C		
PUR-symbol	[weight %]	M_n [g/mol]	M_w [g/mol]	[η] [dl/g]
T/BH	100	26700	33700	0,190
T/BA20/BH(1:4)*	53	39000	51000	0,390
T/BA20/BD(1:4)	38	50490	73240	0,266
T/BA6/BD(1:4)	67	72600	145000	0,363
M/BA20/BD(1:4)	43	53800	142000	0,371
M/BA6/BD(1:4)	76	53300	115000	0,358

*Ratio PBA:BHHBP (mol).

Characterization Techniques

The molecular weights were measured by SEC with the following equipment: Knauer HPLC Pump 64; linear column, JORDI-GEL, mixed bed, 600 × 10 mm i.d., 100 μl of a 0.5% polymer solution in DMF, flow rate 1 ml/min; Differential Refractometer-Viscometer (Knauer Viscotek) for detection. The measurements were made at room temperature. For the calibration of the system a poly(ester urethane) was used. Absolute values of molecular weight and number average M_n were determined by light-scattering and membrane osmometry respectively. The reported inherent viscosities are the results of measurements in the SEC equipment at room temperature.

A Perkin-Elmer DSC 7 instrument was used to determine the thermal transitions at a scanning rate of 20 K/min. The maxima or minima of the endothermic or exothermic peaks were defined as phase transition temperatures. Glass transition temperature T_g was taken from the half-height of the change in heat capacity. Usually the second heating curves and the first cooling curves were used for interpretation. Crystallisation of the PBA 2000 soft segment was observed only in the first scan.

The morphology of the liquid crystalline polyurethanes and blends was studied by electron microscopy with a Tesla TEM 500 at 90 kV. Thin samples were prepared by casting a 1 wt-% polymer solution on freshly cleaved rock salt surfaces. The samples were stained with ruthenium tetroxide. Comparisons with ultrathin cuts showed no differences to the cast specimen.

Tensile tests were performed on a Rauenstein Tiratest 2160 machine.

The 0.1 mm thick films were cut into strips of 100 × 10 mm^2 and measured with a draw rate of 100 mm/min. The test conditions are in accordance with DIN 53455.

RESULTS AND DISCUSSION

Properties of the Unblended Polyurethanes

The composition and molecular weights of the two investigated liquid crystalline polyurethanes are summarized in Table 1.

LCPU is based only on hard segments composed of 2,4-tolylene diisocyanate (T) and the mesogenic unit BHHBP (BH). It is a brittle material with a low solubility in organic solvents. Therefore, the preparation of self-supporting films is complicated.

The DSC thermogram of LCPU was already reported by Stenhouse et al. [29]. Our sample exhibits the following DSC features [Figure 1(a)].

The heating scans show a well resolved step corresponding to the glass transition (T_g) at about 79–84°C. The T_g-region is followed by a broad exotherm, referring to recrystallization with a peak maximum at 120°C and two sharp endotherms appearing at 157°C ($T_{m,1}$) and 169°C ($T_{m,2}$). The total heat of fusion ΔH_m from 145 to 175°C is 45.9 J/g. As reported by Smyth et al. [40,41] and Papadimitrakopulos et al. [42] both endothermic peaks are associated with crystal melting and not with the crystal-mesophase and mesophase-isotropic transitions, respectively. Upon cooling, a sharp exotherm at around 123°C (the crystallization enthalpy ΔHc from 117 to 130°C is 17,1 J/g) and an exothermic tail were observed. The higher temperature exotherm is associated with the isotropic-mesophase transition, the lower one appearing as an exothermic tail is associated with the mesophase-crystal transition [29,40]. The measured phase transition temperatures and enthalpies correspond with values published by Stenhouse et al. [29].

SLCPU is a segmented liquid crystalline polyurethane containing the same hard segment as LCPU but with a polyester soft segment based on poly(butylene adipate) M_n 2000 (BA20). The molar ratio of polyester to BHHBP is 1:4.

The incorporation of the polyester soft segment PBA 2000 does not change the general shape of the DSC pattern [Figure 1(b)]. SLCPU shows a sharp glass transition at about −34°C. The melting peak of the crystallized polyester is in evidence only in the first heating scan [which is not present in Figure 1(b)] at about 50°C (ΔH_m from 36 to

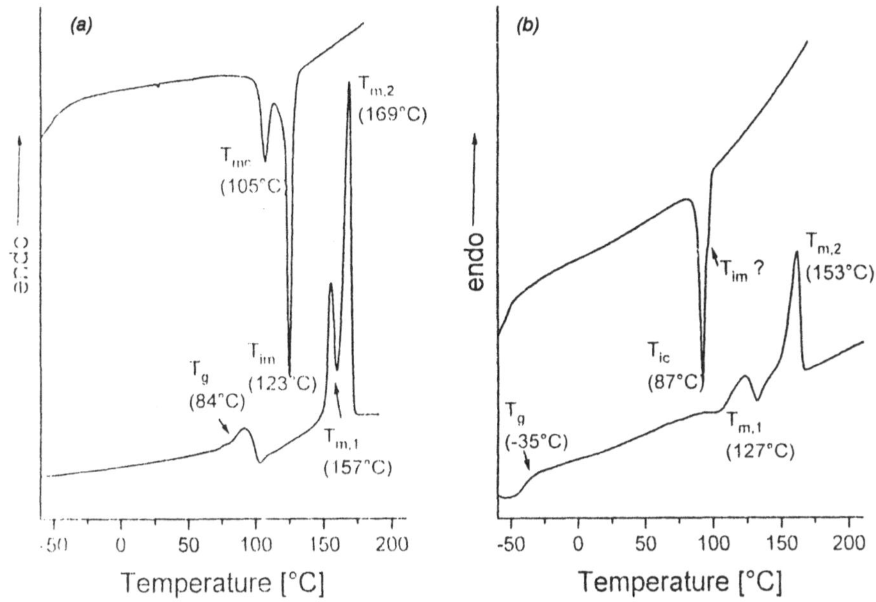

Figure 1. DSC thermograms of LCPU (a) and SLCPU (b).

71 °C is 1.6 J/g). The two endothermic peaks of the mesogenic hard segment appear at 127 °C ($T_{m,1}$) and 153 °C ($T_{m,2}$) in the second scan. The corresponding ΔH_m-values from 98 to 132 °C and from 139 to 162 °C are 5.3 and 7.6 J/g, respectively. On cooling, only one sharp asymmetrical peak at 87 °C is observed. The total crystallization ΔH_c from 78 to 195 °C is 11.0 J/g.

DSC investigations of the six conventional segmented polyurethanes mentioned in Table 1 exhibit glass transition temperatures listed in Figures 2(a) and 2(b). The thermal behaviour of the TDI system is quite different from that of the MDI systems.

The TDI polyurethanes with polyester soft segments and butanediol-1,4 as chain extenders show a higher glass transition temperature at similar hard segment content than the MDI polyurethanes and an enhanced increase of T_g with increasing hard segment content (Table 2). This is caused by a stronger interaction between soft and hard segments and can be interpreted as compatibility between the soft segments and the TDI-based hard segments. Transition temperatures which can be assigned to the hard segment domains are not observed in the heating scans of the TDI polyurethanes, i.e., phase separation of

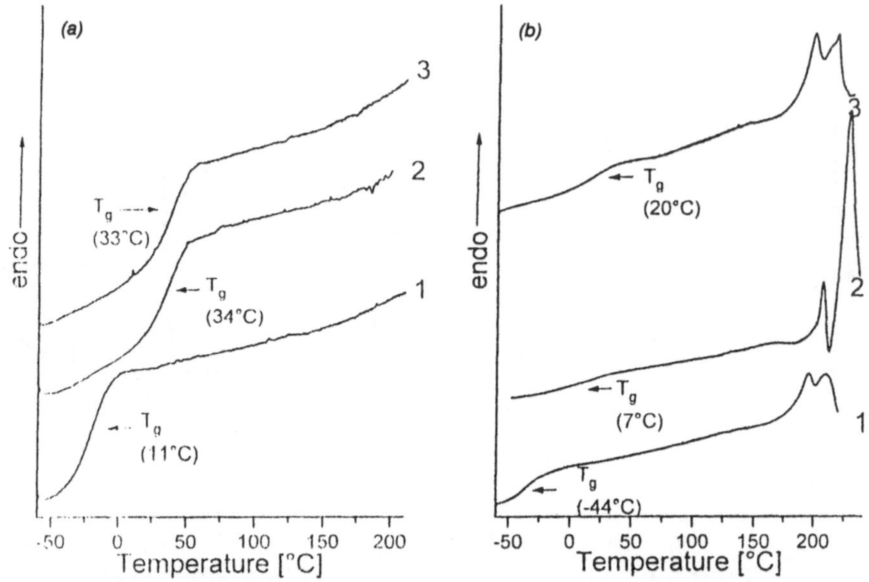

Figure 2. DSC thermograms of conventional polyurethanes, 2nd heating scan: (a) 1:T/BA20/BD(1:4), 2: T/BA6/BD(1:4), 3: T/TH6/BD(1:4); (b) 1:M/BA20/BD(1:4), 2: M/BA6/BD(1:4), 3:M/TH6/BD(1:4).

soft and hard segments does not exist in the TDI systems. This can be of interest in the further studies on blending behaviour.

In contrast to that, MDI polyurethanes exhibit phase separation of soft and hard segments. T_g of the M/BA20 system with nearly the same hard segment content is much lower than T_g of the T/BA20 system. This is caused by the more pronounced phase separation of soft and hard segments in the MDI system. In addition the phase separated hard segments of M/BA20/BD(1:4) lead to endothermic peaks in the region of

Table 2. Glass transition temperatures of the conventional polyurethanes by DSC.

Polyurethane	T_g [°C]
T/BA20/BD(1:4)	−11
T/BA6/BD(1:4)	+34
T/TH6/BD(1:4)	+33
M/Ba20/BD(1:4)	−44
M/BA6/BD(1:4)	+7
M/TH6/BD(1:4)	+22

185–225 °C caused by the melting of different types of microcrystalline hard segment domains. Shortening of the polyester segment length by incorporation of poly(butylene adipate) M_n 600 (BA6) instead of BA20 in the chain influences the temperature and peak shape of the hard segment endothermic transition temperatures and increases the T_g of the short soft segments. This shift may be explained either by partial miscibility [43] or by soft segment mobility reduction by the stiff hard segments.

The introduction of the mesogenic unit BHHBP into the TDI polyurethane instead of butanediol-1,4 promotes phase separation of the hard segments and the formation of microcrystalline hard segment domains. The comparison of the glass transition temperatures in Figure 3 shows that the difference of T_g between T/BA20/BD(1:4) and SLCPU which have similar molecular structures is about 25 °C. This can be explained by a stronger interaction among the mesogenic hard

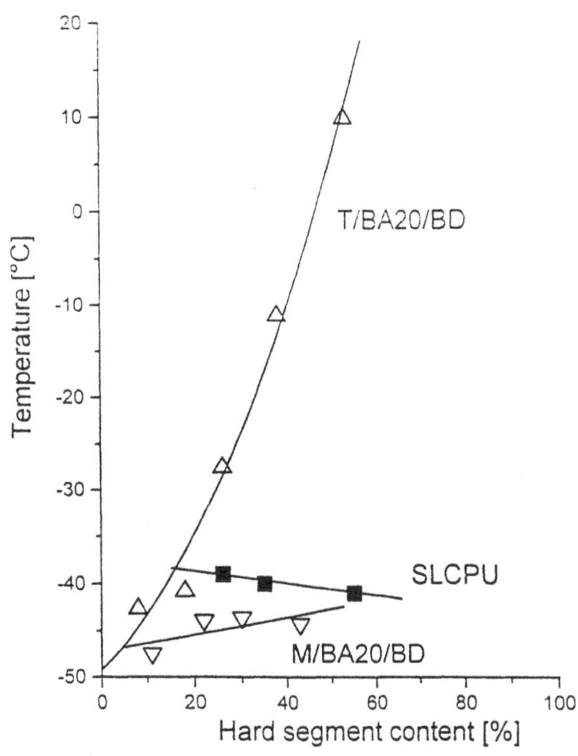

Figure 3. Soft segment-T_gs of SLCPU in comparison with conventional polyurethanes.

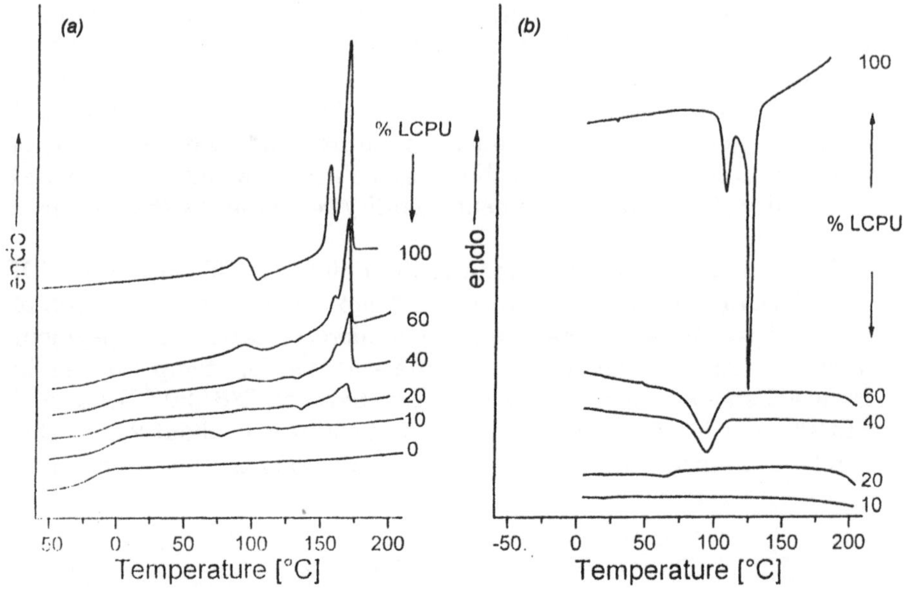

Figure 4. DSC thermograms of blends of LCPU with TDI-polyurethane T/BA20/BD(1:4): (a) 2nd heating scans, (b) 1st cooling scans.

segments in the liquid crystalline polyurethane SLCPU causing phase separation between soft and hard segments. A similar behaviour is observed with the conventional MDI polyurethane M/BA20/BD(1:4).

Blends of LCPU with Conventional Polyurethanes

The DSC diagrams in Figures 4 and 6 show the transition temperatures of mixtures of the nonsegmented liquid crystalline polyurethane LCPU with a conventional TDI [T/BA20/BD(1:4)] or MDI polyurethane [M/BA20/BD(1:4)], respectively. These conventional polyurethanes differ in morphology. In the TDI system a soft and hard segment compatibility or miscibility was observed in contrast to the MDI system with a clear immiscibility of these segments.

The addition of increasing amounts of LCPU to both conventional polyurethanes results in characteristic changes of the DSC diagrams. At a concentration of 20 wt-% LCPU the endothermic peak $T_{m,2}$ at 169 °C is observed [Figure 4(a)] which demonstrates that the mixture has not influenced the crystallisation behaviour of the LCPU. In the mixture with 40 wt-% LCPU the second characteristic endothermic peak $T_{m,1}$ appears at 157°C. The unchanged positions of the two endothermic peaks lead to the conclusion that the liquid crystalline properties of LCPU can be preserved in this blend and that LCPU is immiscible with the investigated TDI polyurethanes.

These results are confirmed by the second cooling scans in Figure 4(b). Despite the immiscibility of LCPU and TDI polyurethanes an interaction between the two polymers exists which influences the ordering process in the LCPU. Only one broad exothermic peak at about 90°C is observed in the mixture with 40 wt-% T/BA20/BD(1:4), whereas the blends with 40 wt-% T/BA6/BD(1:4) (Figure 5) show two

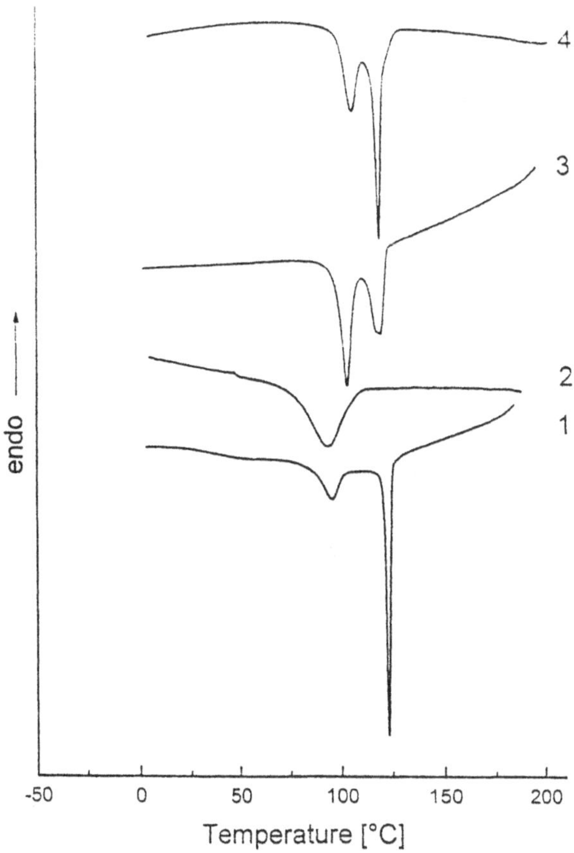

Figure 5. DSC thermograms of blends of 60% LCPU with conventional TDI-polyurethanes, 1st cooling scan: (1) T/BA6/BD(1:4); (2) T/BA20/BD(1:4); (3) T/BA6; (4) T/BA20.

exothermic peaks, which may be interpreted as isotropic to mesophase transition temperature T_{im} at 123°C and mesophase to crystalline transition temperature T_{mc} at about 100°C. The intensity of these peaks is strongly influenced by the structure of the TDI polyurethanes as observed in the mixture with a polyurethane based on TDI and PBA 2000 without butanediol as chain extender. This interesting aspect and its explanation are currently under investigation.

A similar mixing behavior was observed with the MDI polyurethane M/BA20/BD(1:4) as presented in Figure 6(a). The first sign for the existence of LCPU in the mixture is obtained at a LCPU content of 20 wt-% [Figure 6(a)].

The results of blending of LCPU with the conventional polyurethanes mentioned in Table 1 are summarized in Figure 7. Only the MDI polyurethane M/BA6/BD(1:4) differs strongly in the mixing behaviour. This is manifested in the values of the glass transition temperatures of the blends. With increasing content of LCPU the glass transition temperatures T_g of these blends increase significantly. No or very low increases of T_g with growing LCPU content have been observed in LCPU mixtures with the five other polyurethanes. This

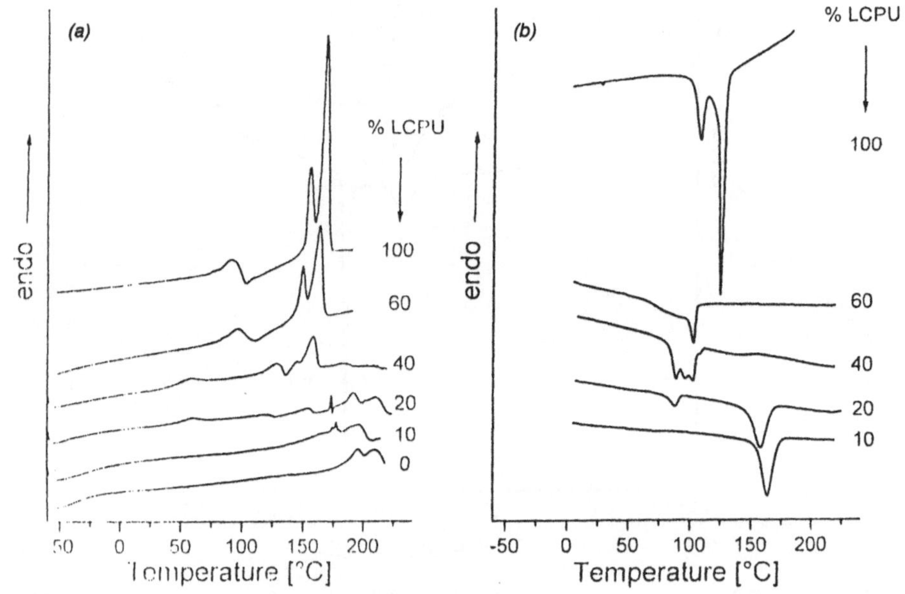

Figure 6. DSC thermograms of blends of LCPU with MDI-polyurethane M/BA20/BD(1:4): (a) 2nd heating scans, (b) 1st cooling scans.

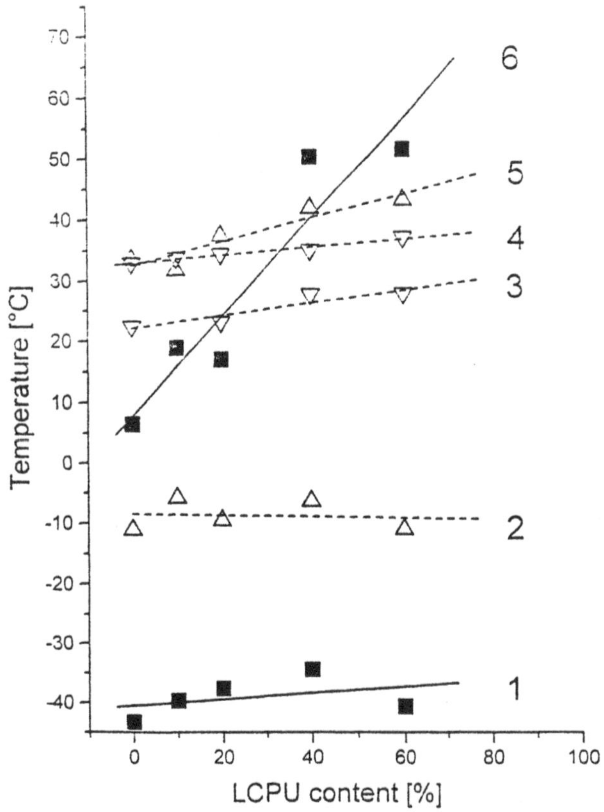

Figure 7. *Glass transition temperatures of blends of LCPU with conventional polyurethanes by DSC: (1) M/BA20/BD(1:4); (2) T/BA20/BD(1:4); (3) M/TH6/BD(1:4); (4) T/TH6/BD(1:4), (5) T/BA6/BD(1:4), (6) M/BA6/BD(1:4).*

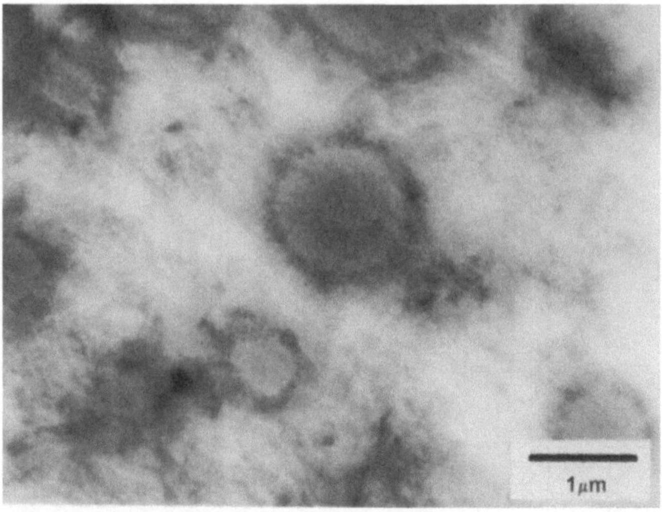

(a)

(b)

Figure 8. *Electron micrographs of blends of LCPU with conventional polyurethanes: (a) T/BA6/BD(1:4), (b) M/BA6/BD(1:4).*

result proves a higher interaction between LCPU and M/BA6/BD(1:4) and may be explained by partial miscibility of these two polyurethanes. It is assumed that the LCPU is partly incorporated in the soft segment phase of the MDI polyurethane.

These conclusions from DSC measurements–immiscibility and partial miscibility–are in accordance with electron microscopic observations [Figures 8(a) and (b)] which exhibit pronounced phase separation in the LCPU-T/BA6/BD(1:4) blends compared to the mixtures with M/BA6/BD(1:4).

Blends of SLCPU with Conventional Polyurethanes

SLCPU is characterized by phase separation between soft and hard segments in contrast to the analogous polyurethane with a nonmesogenic hard segment based on 1,4-butanediol. The question is whether the two identical soft segments BA20 will mix without influencing the liquid crystalline behaviour of the SLCPU.

The DSC diagrams of SLCPU mixtures with the conventional TDI polyurethanes T/BA20/BD(1:4) and T/BA6/BD(1:4) are shown in

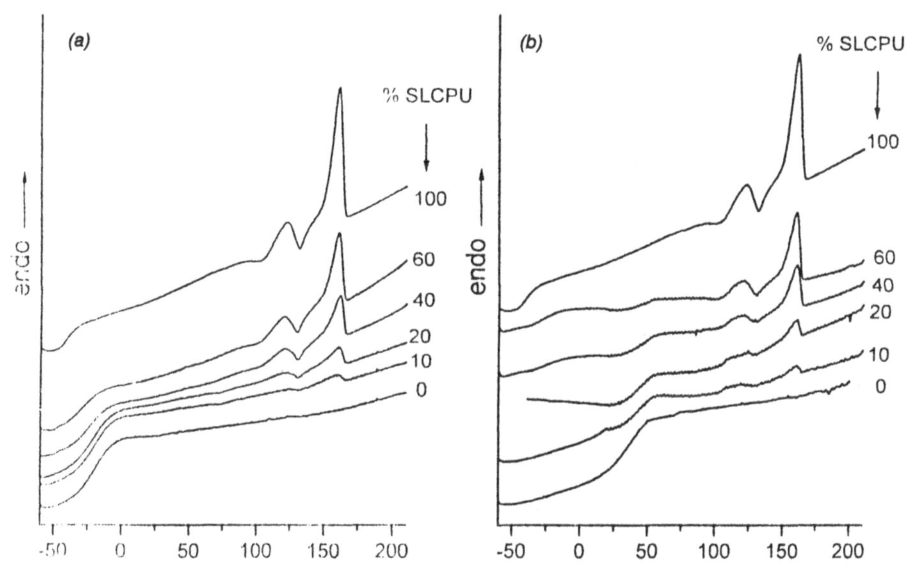

Figure 9. DSC thermograms of blends of SLCPU with conventional TDI-polyurethanes: (a) T/BA20/BD(1:4), (b) T/BA6/BD(1:4).

Figures 9(a) and 9(b). The first traces of the endothermic peaks of SLCPU are observed in the second heating scan of the T/BA20/BD(1:4) blend with 10 wt-% of SLCPU [Figure 9(a)]. Similar results for hard segment behaviour are obtained in the mixtures of SLCPU with T/BA6/BD(1:4) as can be seen in Figure 9(b).

A different behaviour, however, is observed for the glass transition temperature of the blends. The results are summarized in Figure 10. The blends with T/BA20/BD(1:4) show typical miscibility behaviour of the soft segments with increasing SLCPU content. The T_g of the polyurethane T/BA20/BD(1:4) at −20°C is provoked by the incorporation of T/BD hard segments in the polyester soft segments. The addition of further soft segments from SLCPU increases the ratio of polyester soft segments to T/BD hard segments in this mixture and causes the decrease in T_g. It seems that the LC hard segment segregation is not influenced by the compatibility of the soft segments.

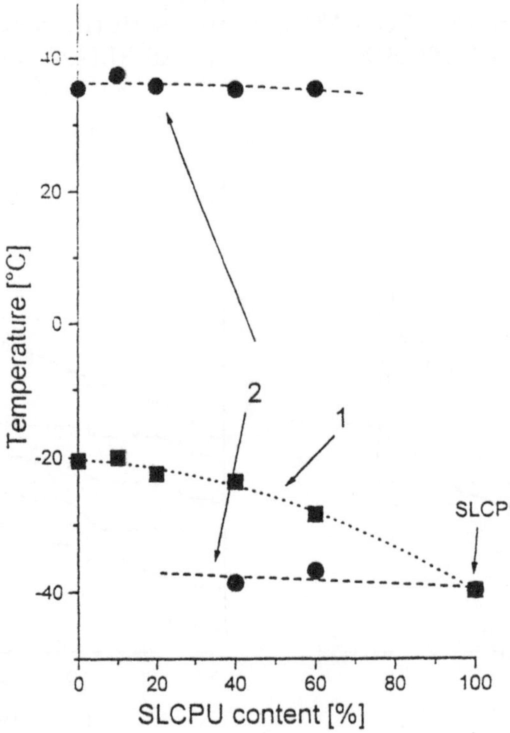

Figure 10. *Glass transition temperatures of blends of SLCPU with conventional TDI-polyurethanes by DSC: (1) T/BA20/BD(1:4); (2) T/BA6/BD(1:4).*

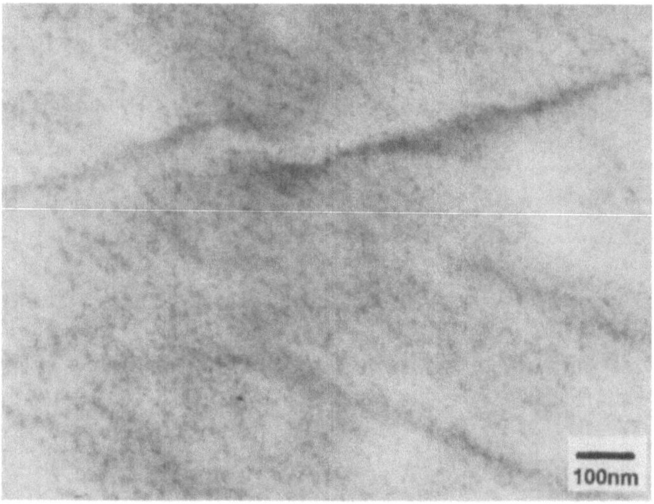

(a)

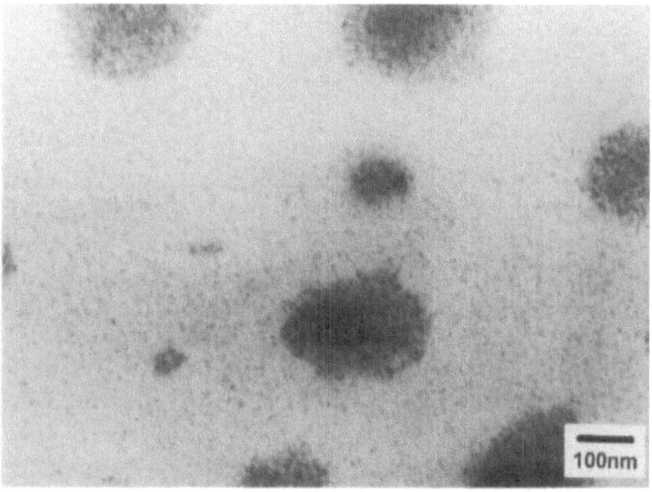

(b)

Figure 11. *Electron micrographs of blends of SLCPU with conventional TDI-polyurethanes: (a) T/BA20/BD(1:4), (b) T/BA6/BD(1:4).*

Shortening of the polyester chain length from M_n 2000 (BA20) to 600 (BA6) hinders the miscibility of the soft segments. Two distinct soft segment glass transitions can be observed in the mixture (Figure 10). The transmission electron micrographs clearly show the differences in morphology caused by the different miscibility of the soft segments. The T/BA20-system is less inhomogeneous than the T/BA6-system [Figures 11(a) and 11(b)].

While the interaction of the T/BD hard segments with the mesogenic T/BH hard segments is relatively weak (phase separation), a stronger interaction is observed with M/BD hard segments. Independent of the chain length of BA20 and BA6 in the MDI polyurethanes the two characteristic endothermic peaks of SLCPU are first observed at a concentration of 40 wt-% SLCPU in the mixtures as shown in Figures 12(a) and 12(b). This can be explained by the heat of fusion of the second melting peak (Figure 13). The heat of fusion of SLCPU is unchanged in presence of T/BA20/BD(1:4), whereas a clear decrease of the heat of fusion is observed in the mixture 40:60 of SLCPU:M/BA20/BD(1:4). Analogous results are obtained with blends of SLCPU and M/BA6/BD(1:4). The stronger interaction of the polyurethane mole-

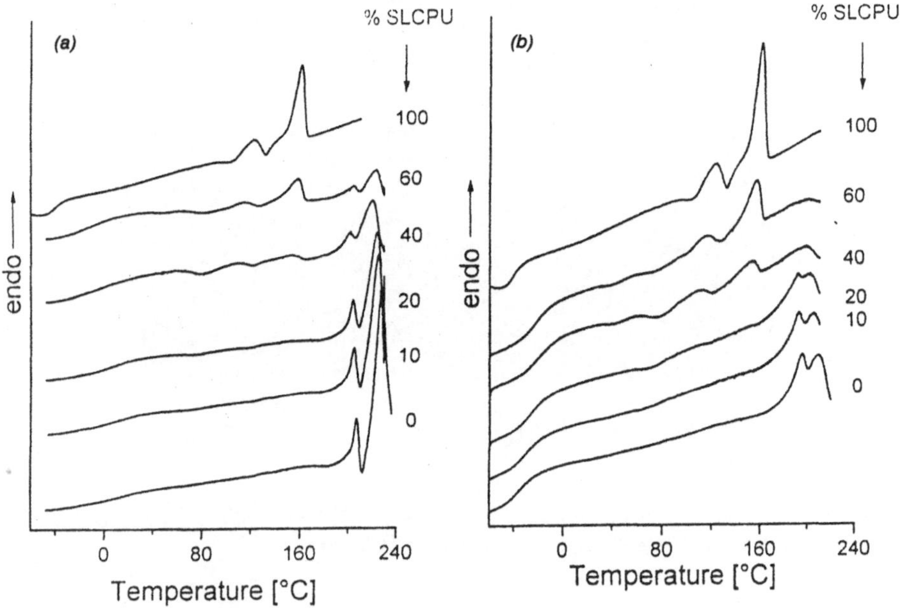

Figure 12. DSC thermograms of blends of SLCPU with conventional MDI-polyurethanes: (a) M/BA20/BD(1:4), (b) M/BA6/BD(1:4).

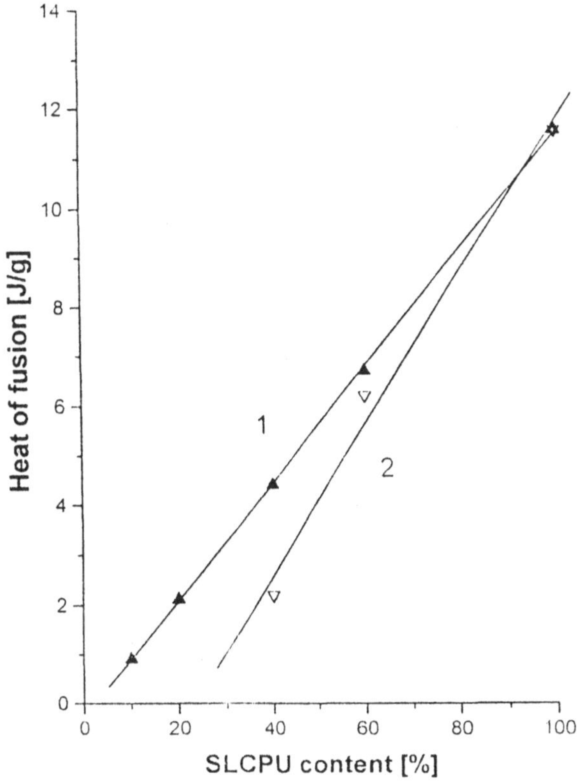

Figure 13. *Heat of fusion of the second melting peak ($T_{m,2}$) of SLCPU in blends with conventional polyurethanes: (1) T/BA20/BD(1:4), (2) M/BA20/BD(1:4).*

cules, particularly of their hard segments, hinders the phase separation of the mesogenic hard segments. In the third heating scan the heat of fusion remains unchanged. Therefore the observations are caused by miscibility and not by transurethanification as discussed by Eisenbach [44] and Joel [45]. The mesophase formation in the investigated blends is controlled by thermodynamic and kinetic processes.

In the blends of SLCPU and M/BA20/BD(1:4), presented in Figure 14, No. 1, the incorporation of SLCPU hard segments in the mixture of the soft segments of SLCPU and M/BA20/BD(1:4) raises the glass transition temperature. Transmission electron micrographs of blends with MDI-polyurethanes with long as well as short soft segments (Figure 15) do not show the coarse inhomogeneities of the TDI-polyurethanes with long soft segments. This observation corresponds with the results from DSC.

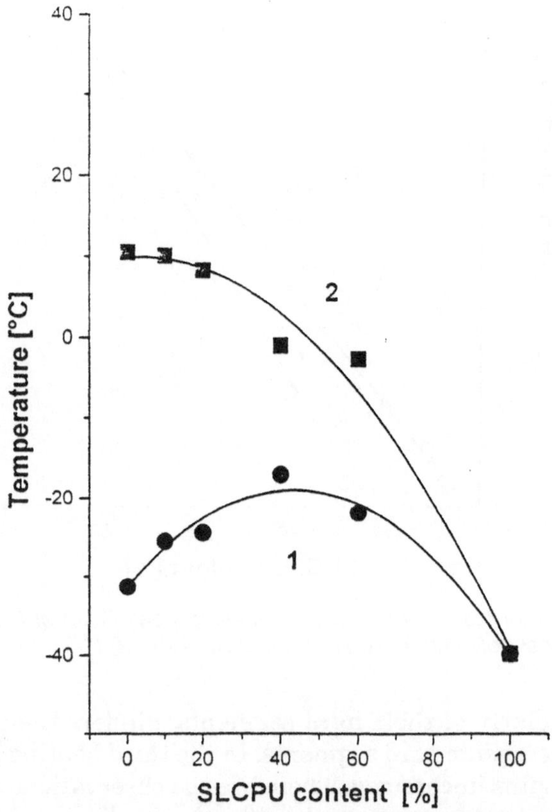

Figure 14. *Glass transition temperatures of blends of SLCPU with conventional MDI-polyurethanes by DSC: (1) M/BA20/BD(1:4), (2) M/BA6/BD(1:4).*

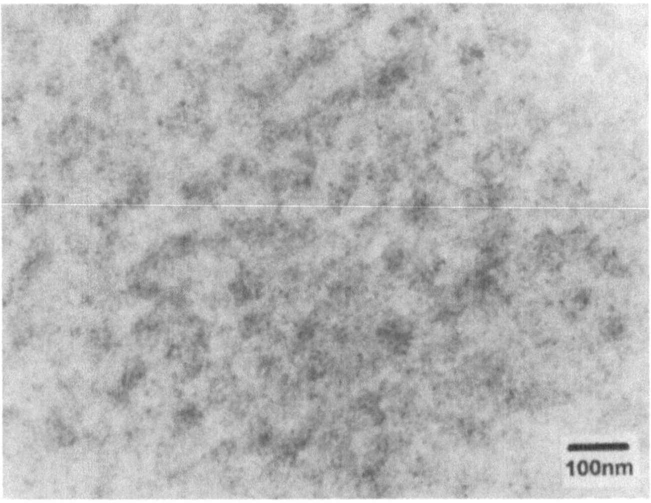

(a)

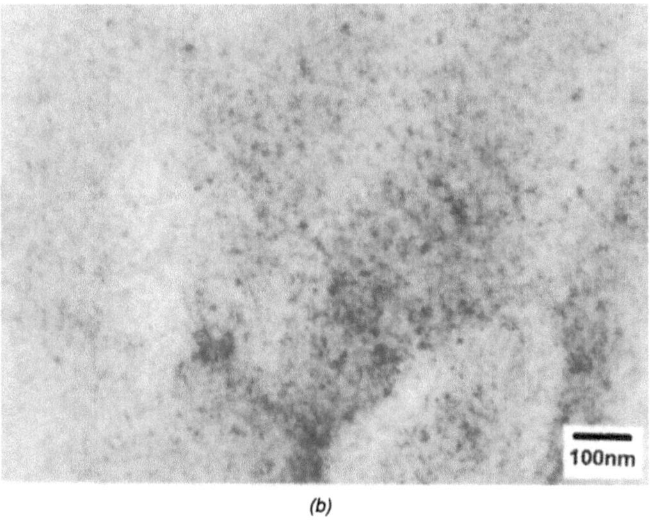

(b)

Figure 15. *Electron micrographs of blends of SLCPU with conventional MDI-polyurethanes: (a) M/BA20/BD(1:4), (b) M/BA6/BD(1:4).*

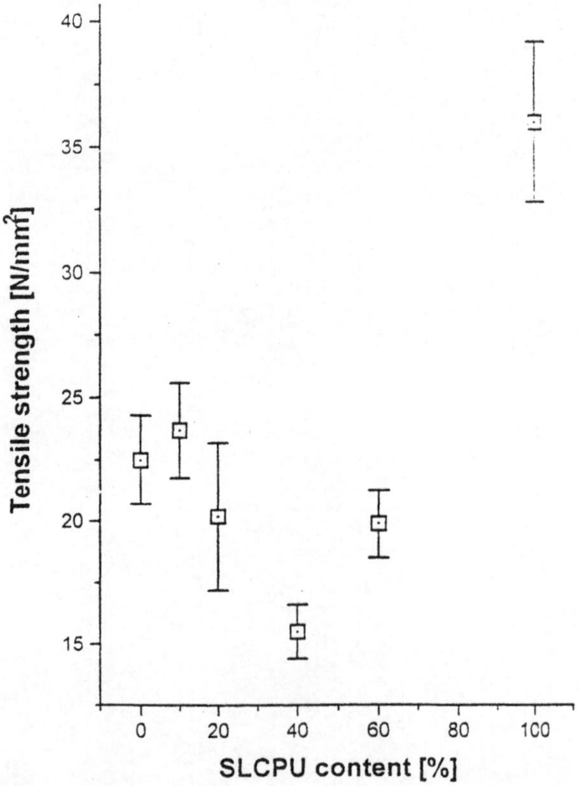

Figure 16. *Tensile strength of blends of SLCPU with MDI-polyurethane M/BA6/BD(1:1).*

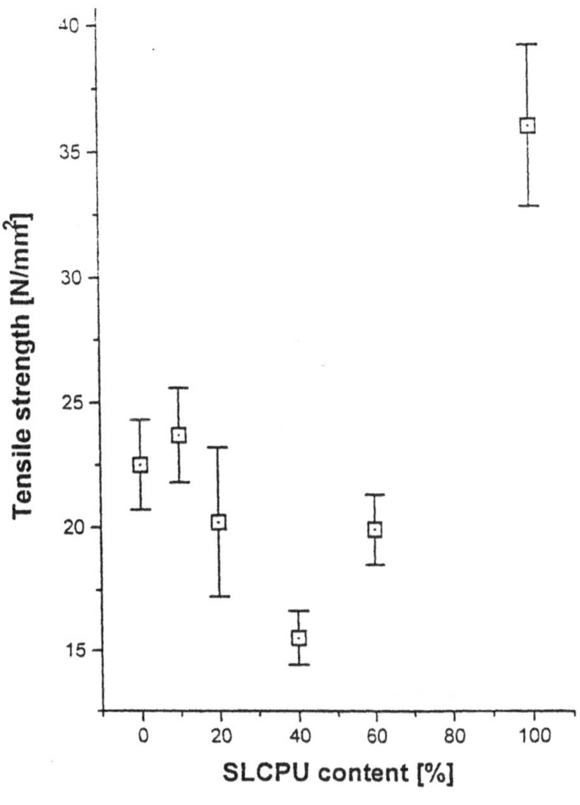

Figure 17. *Tensile strength of blends of SLCPU with MDI-polyurethane M/BA20/BD(1:4).*

For future technological applications it may be of interest that the highly ordered structure in liquid crystalline polyurethanes can be preserved in mixtures with conventional polyurethanes and possibly other polymers. The high tensile strength of the SLCPU is transferable to the blend if phase separation of the mesogenic hard segments can be preserved. For example, this is shown in blends with an MDI-polyurethane with only little hard segment content (Figure 16). In blends, however, in which the M/BD-segment content is high and phase separation of the LCPU hard segments is hindered we do not get this property transfer (Figure 17).

CONCLUSIONS

In most cases the nonsegmented liquid crystalline polyurethane LCPU is immiscible with elastomeric TDI and MDI polyurethanes. A partial miscibility has been observed with a MDI polyurethane based on a short polyester soft segment (PBA 600) and a reasonably high hard segment content.

The incorporation of a polyester soft segment in the LCPU increases the possibilities for partial miscibility with polyester based elastomeric polyurethanes. Independent of that, the liquid crystalline properties can be obtained in blends with TDI polyurethanes, whereas elastomeric MDI polyurethanes partially hinder the mesophase formation.

Partial miscibility of the two polymers and preservation of the LC properties are necessary for transferring the high tensile strength of the LC polyurethane to the blend.

REFERENCES

1. Olabisi, O., L. M. Robeson and M. T. Shaw. 1978. *Polymer-Polymer Miscibility.* New York: Academic Press.
2. Paul, D. R. and S. Newman. 1978. *Polymer Blends.* New York: Academic Press.
3. Klempner, D. and K. C. Frisch. 1983. *Polymer Alloys III.* New York and London: Plenum Press.
4. Jacobson, H. S., D. J. Gordon, G. V. Nelson and A. Balazs. 1992. *Advanced Materials.* 4:198.
5. Siol, W. 1991. *Makromol. Chem., Macromol. Symp.* 44:47.
6. Kammer, H. W. 1986. *Acta Polymerica.* 37:1.

7. Saunders, J. H. and K. C. Frisch. *Polyurethanes, Chemistry and Technology.* 1962, *Part I Chemistry,* 1964, *Part II Technology.* New York: Interscience.
8. Cummings, A. R. C. and P. Wright. 1969. *Solid Polyurethane Elastomers.* London: Maclaren and Sons.
9. *Kunststoff-Handbuch, Band VII, Polyurethane,* 1983. H. Becker, ed., Munich: Carl Hanser Verlag.
10. Eisenbach, C. D., M. Baumgartner and C. Gunter. 1986. *Advances in Elastomers and Rubber Elasticity,* J. Lai and J. E. Mark, eds., New York: Plenum.
11. Eisenbach, C. D. and H. Nefzger. 1989. *Multiphase Macromolecular Systems.* B. M. Culbertson, ed., New York: Plenum.
12. Süling, C. H. et al. 1986. Ger. Offen. 3.628.141, Bayer AG.
13. Tanaka, M. and T. Nakaya. 1993. *Advances in Urethane Science and Technology, Vol. 12.* K. C. Frisch and D. Klempner, eds., Lancaster, Basel: Technomic Publishing Co., Inc.
14. Iimura, K., N. Koide, H. Tanabe and M. Takeda. 1981. *Macromol. Chem.,* 182:2569.
15. Mormann, W. and A. Baharifar. 1990. *Polymer Bull.,* 24:413.
16. Mormann, W. and S. Benadda. 1991. *Macromol. Chem., Macromol. Symp.,* 50:229.
17. Mormann, W. and M. Brahm. 1989. *Macromol. Chem.,* 190:631.
18. Mormann, W. and M. Brahm. 1991. *Macromolecules,* 24:1096.
19. Mormann, W. and M. Brahm. 1990. *Mol. Cryst. Liq. Cryst.,* 185:163.
20. Mormann, W. and S. Benadda. 1991. *Macromol. Chem.,* 192:2411.
21. Mormann, W. 1989. *ACS Polymer Prepr.,* 30:291.
22. Kossmehl, G. and B. Hirsch. 1990. *Acta Polymerica.* 41:597.
23. Tanaka, M. and T. Nakaya. 1986. *Macromol. Chem.,* 187:2345.
24. Tanaka, M. and T. Nakaya. 1987. *J. Macromol. Sci.,* A24:777.
25. Tanaka, M. and T. Nakaya. 1986. *Kobunshi Ronbunshu,* 43:311, 1986 Chem. Absrt. 105:79634c.
26. Lorenz, R., M. Els, F. Haulena and O. Lorenz. 1990. *Angew. Macromol. Chem.,* 180:51.
27. MacKnight, W. J. and F. Papadimitrakopoulos. 1993. *Macromol. Chem., Macromol. Symp.,* 69:41.
28. Pollack, S. K., G. Smyth, F. Papadimitrakopoulos, P. J. Stenhouse, S. L. Hsu and W. J. MacKnight. 1992. *Macromolecules,* 25:2381.
29. Stenhouse, P. J., E. M. Valles, S. W. Kantor and W. J. MacKnight. 1989. *Macromolecules,* 22:1467.
30. Sato, M. F. Komatsu, N. Takeno and K. Mukaido. 1991. *Makromol. Chem. Rapid Comm.,* 12:167.
31. Kricheldorf, H. R. and J. Awe. 1989. *Makromol. Chem.,* 190:2779.
32. Kricheldorf, H. R. and J. Awe. 1989. *Makromol. Chem.,* 190:2797.
33. Penczek, P. K., K. C. Frisch, B. Szczepaniak and E. Rudnik. 1993. *J. Polym. Sci.,* A31:1211.

34. Szczepanaik, B., K. C. Frisch, P. Penczek, J. Mejsner, I. Leszcyhnska and E. Rudnik. 1993. *J. Polym. Sci.,* A31:3223.
35. Szczepaniak, B., K. C. Frisch, P. Penczek, E. Rudnik and M. Cholisnka. 1993. *J. Polym. Sci.,* A31:3231.
36. Tang, W., R. J. Farris, W. J. MacKnight and C. D. Eisenbach. 1994. *Macromolecules,* 27:2814.
37. He, X., X. Jia and X. Yu. 1994. *J. Appl. Polym. Sci.,* 54:207.
38. Mix, R., H. Goering, G. Schulz, E. Gründemann and J. Gähde. 1995. *J. Polym. Sci., Part A, Polymer Chemistry,* in press.
39. Kressler, J. and H. W. Kammer. 1991. *3rd Dresdner Polymer Discussion,* Preprint, p. 56.
40. Smyth, G., F. M. Valles, S. K. Pollack, J. Grabowicz, P. J. Stenhouse, S. L. Hsu and W. J. MacKnight. 1990. *Macromolecules,* 23:3389.
41. Smyth, G., S. K. Pollack, W. J. MacKnight and S. L. Hsu. 1990. *Liquid Crystals,* 7:839.
42. Papadimitrakopoulos, F., S. L. Hsu and W. J. MacKnight. 1992. *Macromolecules,* 25:4672.
43. Chang, W. L., T. Baranowski and T. Karalis. 1994. *J. Appl. Polym. Sci.,* 51:1077.
44. Eisenbach, C. D., H. Nefzger, M. Baumgartner and C. Günther. 1985. *Ber. Bunsen-Ges. Phys. Chem.,* 89:1190.
45. Joel, D. and A. Hauser. 1994. *Angew. Makromol. Chem.,* 217:191.

Liquid Crystalline Polyurethanes

PIOTR PENCZEK AND BARBARA SZCZEPANIAK
Industrial Chemistry Research Institute
01-793 Warsaw, Poland

KURT C. FRISCH
Polymer Institute, University of Detroit Mercy
Detroit, MI 48821, U.S.A.

ALEKSANDRA WOLIŃSKA-GRABCZYK
Institute of Coal Chemistry, Polish Academy of Science
444-101 Gliwice, Poland

ABSTRACT

A broad review of the literature on liquid crystalline polyurethanes and of the results of the authors' research work is given. Liquid crystalline polyurethanes with the mesogenic moieties in the main chain are described. Diisocyanate or aromatic diols form the mesogenic units of the polymer chain. The synthesis of liquid crystalline polyurethanes from 2,4-tolylene diisocyanate, 4,4′-bis(2-hydroxyethoxy)biphenyl and poly(oxytetramethylene)diol as flexible spacer is described in detail. The effects of the process of synthesis (one-stage vs. two-stage), content of the mesogenic diol, content and chain length of the flexible spacer, molecular weight of the polyurethane and annealing on phase transitions were studied. The results of the investigation of liquids transport properties of membranes made of the liquid crystalline polyurethanes are presented. The membranes are suitable for the separation of organic liquids.

INTRODUCTION

NUMEROUS POLYMERS EXHIBITING LIQUID CRYSTALLINE PROPERTIES have been investigated. The first liquid crystalline polymers involved polyesters, polyacrylates, polymethacrylates, silicones and many others [1–3].

Liquid crystalline polymers are synthesized from specific monomers containing mesogenic groups. These groups tend to form crystalline phases. The mesogenic moieties are contained in the main chain or in the side chains as pendant groups.

Liquid crystalline polyurethanes have been synthesized by incorporating the rigid aromatic ring moieties into the main chain or into the side chains. A high concentration of the thus built-in mesogenic moieties results in high melting points, usually exceeding the decomposition temperature of the polyurethane. Moreover, such polyurethanes exhibit a limited solubility. This is an effect of high molecular cohesion and low mixing entropy of the liquid crystalline polymers.

To decrease the phase transition temperatures, to broaden the temperature range of the mesophase and to improve the solubility, flexible chain spacers are inserted between rigid mesogenic segments in the main chain or between the pendant mesogenic groups and the main chain of the polyurethane.

Other commonly used methods consist in incorporating of a pendant flexible chain or a bulky substituent as well as in the introduction of unsymmetric comonomers, which disrupts the linearity of the molecule.

Before the first liquid crystalline urethane polymers were described, low molecular weight carbamates had been synthesized. Substituted phenyl carbamates of cholesterol and other steroids were found to exhibit liquid crystalline properties [4]. Diurethanes with the *p*-terphenyl mesogenic moiety (1) show smectic mesophases, whereas a bis-terephthalate (2) exhibits nematic crystallinity [5].

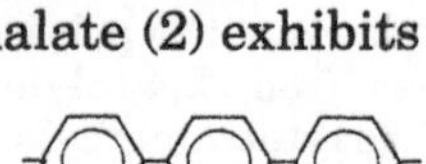

(1)

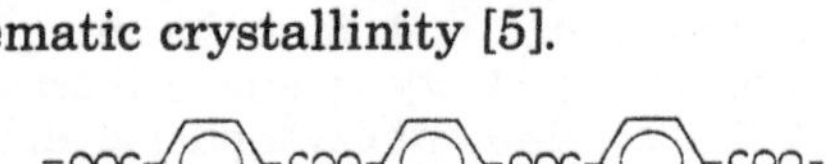

(2)

Research on liquid crystalline polyurethanes started in the 1980s. Nevertheless they did not appear in books devoted to liquid crystalline polymers [1–3].

Liquid crystalline polyurethanes were described first by Iimura et al. [6]. As the mesogenic component, 3,3′-dimethyl-4,4′-biphenyldiyl diisocyanate (Table 1) was used. Oligoethylene glycols (di-, tri-, and tetraethylene glycols) were applied as the diols, playing also the role of flexible spacers. Aliphatic α,ω-alkanediols (2–12 C atoms) were also used. Only thermal properties of the polyurethanes were reported. In a similar series of experiments, 4,4′-azobenzene diisocyanate (from 4,4′-diaminoazobenzene and phosgene) was used [7]. Liquid crystalline properties were found only when using triethylene glycol. A nematic phase exists above 225 °C.

Table 1. Main chain mesogenic components of liquid crystalline polyurethanes.

Mesogenic Moiety	Mesogenic Compound or Polymer	Reference
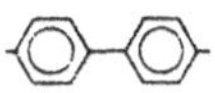	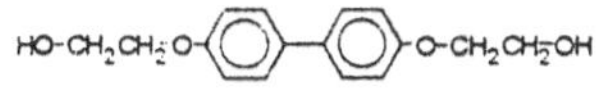	[11, 20, 28, 36]
	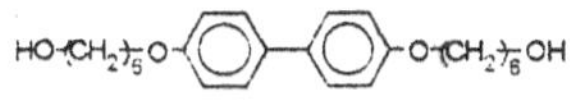	[27, 30-33, 35]
	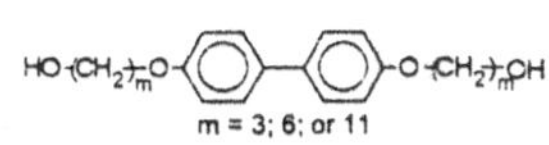	[22]
	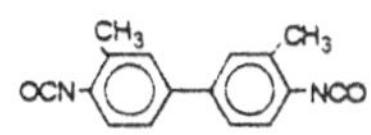	[6]
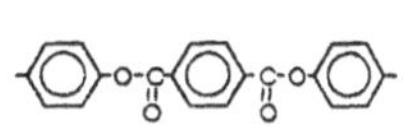	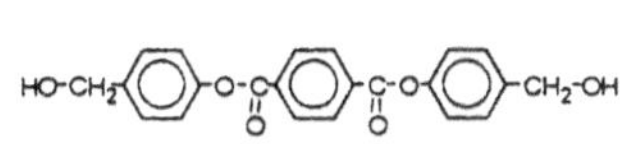	[8, 13]
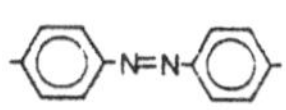	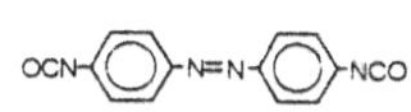	[7]
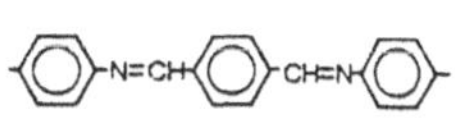	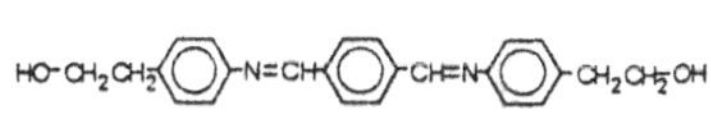	[12]
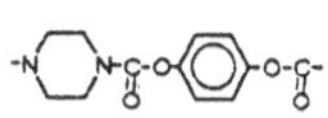	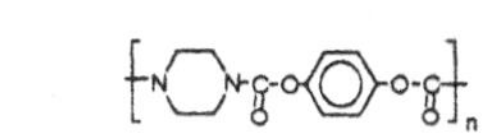	[17]
	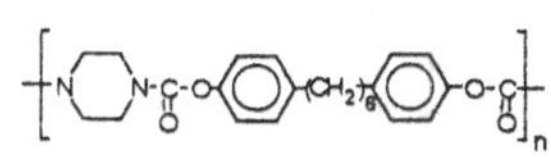	[17]
	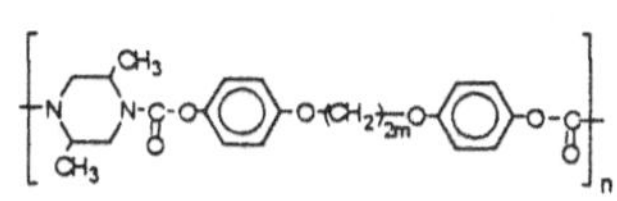	[18]
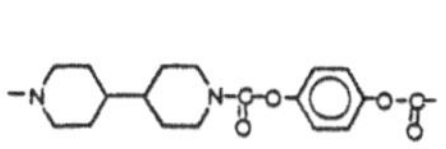	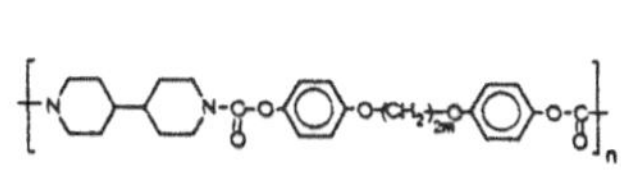	[18]
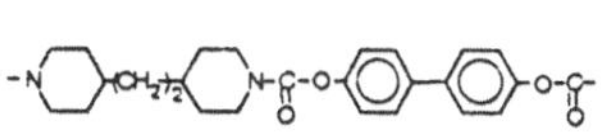	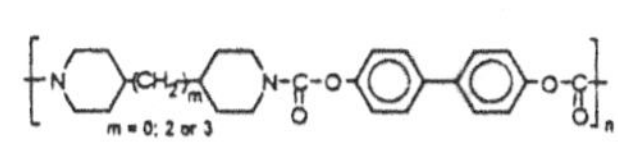	[19]

Then articles by other Japanese authors appeared [8–10]. They used the mesogenic diols presented in Table 1 and synthesized various main-chain thermotropic polyurethanes.

The starting mesogenic diols were synthesized as follows:

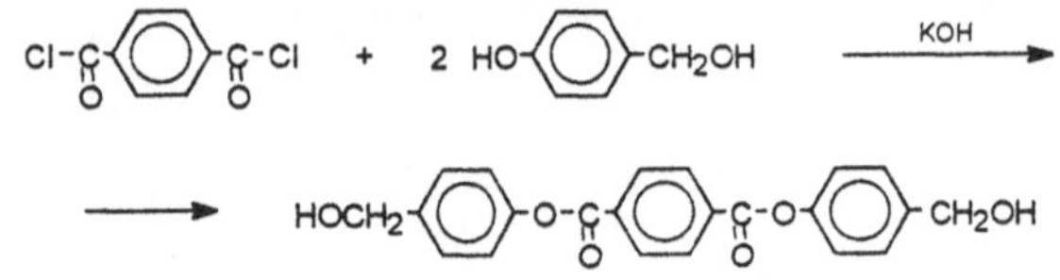

(3)

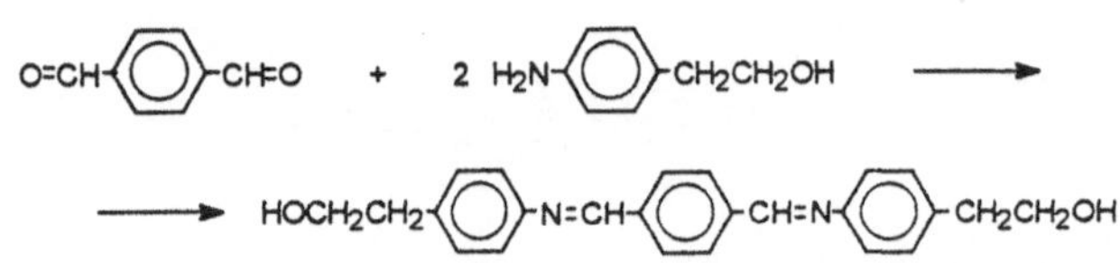

(4)

Several authors compared the phase transitions in polyurethanes synthesized from a mesogenic diol and various aliphatic, cycloaliphatic and aromatic diisocyanates [8,9,11–16]. The following commercially available diisocyanates were used: hexamethylene diisocyanate (HDI), hydrogenated MDI ($H_{12}MDI$), hydrogenated m-xylylene diisocyanate (H_6XDI), 2,4-tolylene diisocyanate (TDI), 4,4′-diphenylmethane diisocyanate (MDI) and 1,4-phenylene diisocyanate (PPDI). In some cases only hexamethylene diisocyanate gave liquid crystalline polyurethanes. In another series of experiments, liquid crystalline properties were obtained with all investigated isocyanates except MDI [12].

A different structure of linear liquid crystalline polyurethanes is presented in Table 1 [17–19]. These polyurethanes were synthesized by means of an isocyanate-free method. Piperazine or bipiperidine derivatives were reacted with 1,4-phenylene or 4,4′-biphenyldiyl di(chloroformates). Polyurethanes containing no N–H groups which would be capable of hydrogen bond formation were thus obtained:

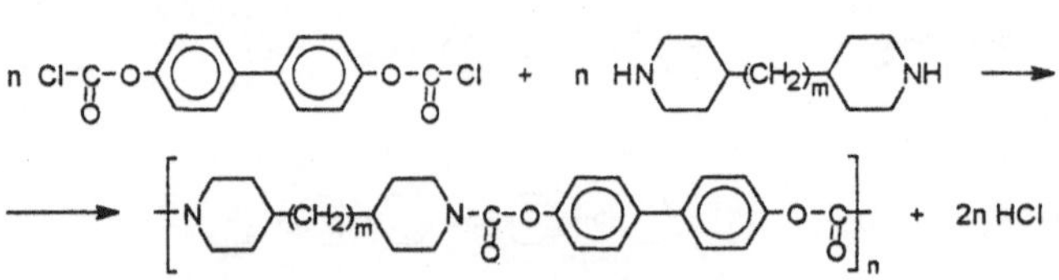

(5)

Besides the diisocyanate component, mesogenic diols also affect the transition temperatures. The influence of 2-hydroxyethoxy and

6-hydroxyhexoxy derivatives of biphenyl, azobenzene and benzophenone was compared [15]. The starting diols can be easily obtained by the reaction of diphenols, e.g., 4,4′-dihydroxybiphenyl, with ω-chloroalkanes, e.g., 6-chlorohexanol:

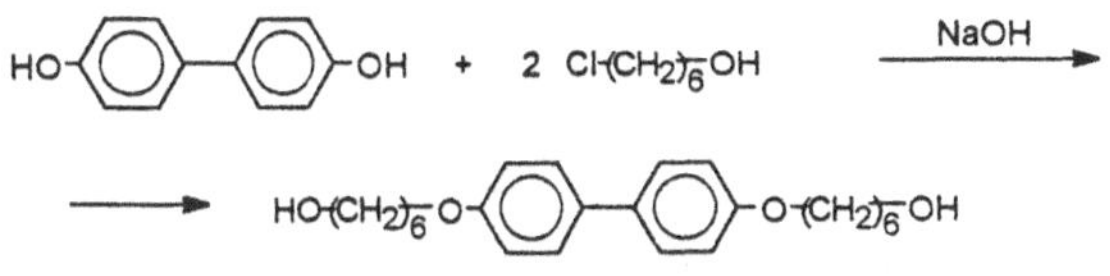

(6)

The authors used the reaction with ethylene carbonate to obtain a 2-hydroxyethoxy derivative [20].

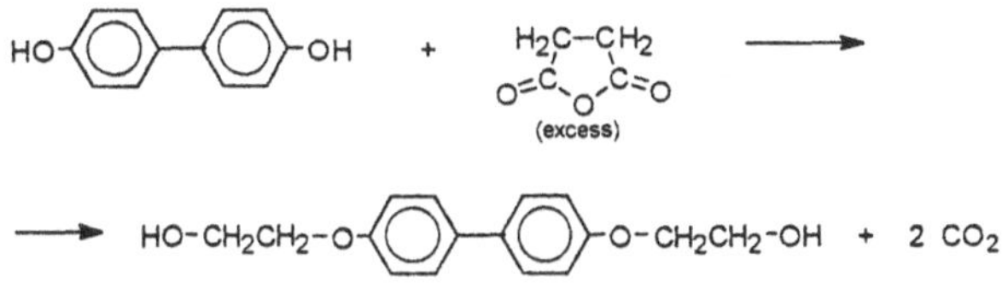

(7)

Segmented polyurethanes with liquid crystalline polyester hard segments were investigated [21].

Liquid crystalline urethane ionomers were synthesized using ionic chain extenders [22]. The polymers are water emulsifiable.

Amphiphilic polyurethanes containing hydrophobic long alkyl groups in the side chains and amino or quaternary ammonium groups in the main chain exhibit smectic structure (Table 2) [23,24]. Cholesteryl groups were built into a side chain of an amphiphilic polyurethane thus imparting liquid crystalline properties (Table 2) [24,25].

A simple mesogenic moiety can appear in the side groups. The azobenzene group was connected with the polyurethane main chain through a flexible polymethylene spacer (Table 2) [14]. Various diisocyanates were used. Only polyurethanes synthesized using hexamethylene diisocyanate exhibited a mesomorphic phase.

Recently, liquid crystalline polyurethanes with mesogenic moieties in both the main chain and side chains were described [25a]. They were synthesized by the polyaddition of hexamethylene diisocyanate with mixtures of two bisphenols:

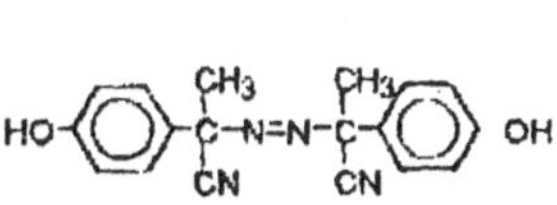

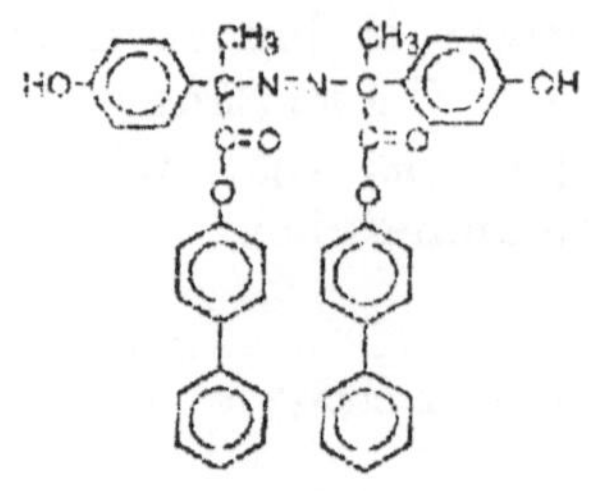

In contrast to the syntheses of liquid crystalline polyurethanes described so far, the polyaddition with diisocyanates was carried out using bisphenols and not their hydroxyalkyl derivatives. Polyurethanes exhibiting relatively low molecular weight values (M_n and M_w below 7,000) were thus obtained.

The polyurethanes exhibited a broad temperature range of mesophase (from 58 to 144 °C). A monotropic mesophase was evidenced by the microscopic observations.

A different approach to the synthesis of liquid crystalline polyurethanes consists in the polycondensation of a mesogenic diol

Table 2. Side chain (pendant) mesogenic components of liquid crystalline polyurethanes.

Mesogenic Moiety	Reference
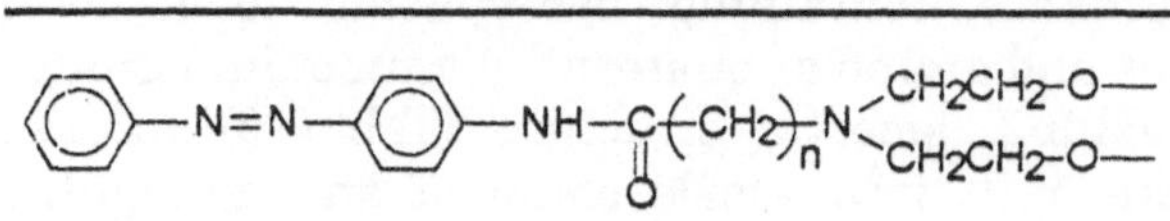	[14]
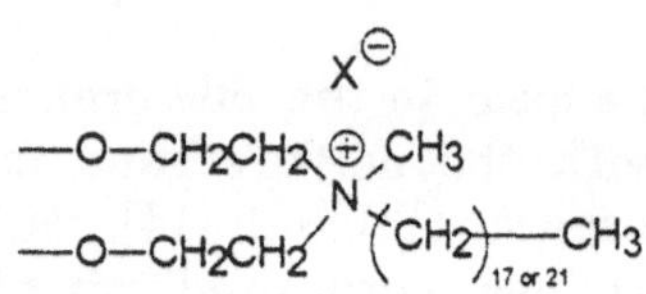	[24, 25]
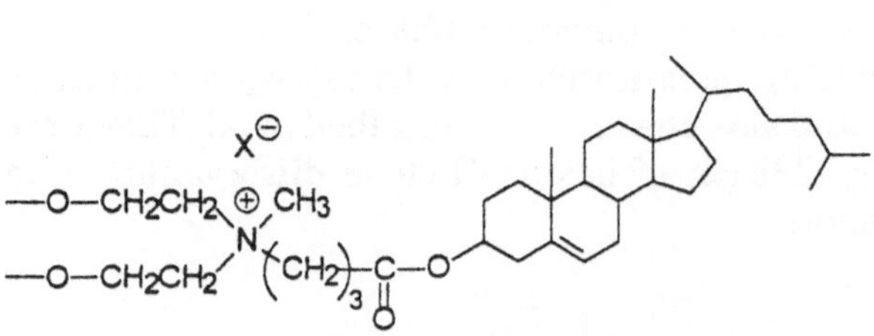 cholesteryl	[23]

with an alkylene bis(phenyl carbamate) with phenol abstraction [26,27]. The details are given in the section entitled "Preparation."

Flexible spacers are contained in the molecule of a mesogenic diol, e.g., hexamethylene segments in 4,4′-bis(6-hydroxyhexoxy)biphenyl (Table 1), which was used by many authors. However, a separate flexible diol is usually built in as spacer, the most popular one being poly(oxytetramethylene)diol PTMO [15,20,28].

Urethane elastomers were recently synthesized from the mesogenic unit

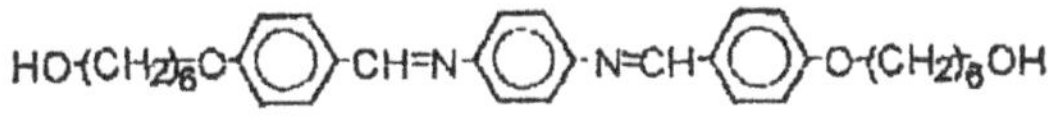

hexamethylene diisocyanate and 1,10-decanediol or PTMO as flexible spacers [28a]. The mechanical properties of cast films were reported.

In recent publications more attention is paid to properties of liquid crystalline polyurethanes. Polyaddition of 2,4-TDI with alkylene di[4-(ω-hydroxyalkoxy-4-oxybenzoyl)oxy-benzoate]s resulted in thermotropic urethane polymers with one or two "cybotactic" nematic mesophases [29]. The phase structure (soft and liquid crystalline hard segments) of segmented polyurethanes synthesized from 2,4-TDI, poly(1,4-butylene adipate)diols with various molecular weights and a mesogenic diol were investigated [30]. The miscibility of similar polyurethanes with common segmented polyurethanes was studied [31].

Phase transitions and crystalline structure were investigated for polyurethanes derived from 2,4-TDI and 4,4′-bis(6-hydroxyhexoxy)biphenyl [32–35]. The low temperature peak, which was related to the crystalline melting transition, depended on the molecular weight and the thermal history whereas the high temperature peak was independent from the clearing transition [32]. A conformationally disordered crystal role of hydrogen bonding was stressed [34].

PREPARATION

Liquid crystalline polyurethanes are usually synthesized by the polyaddition of a diisocyanate with a mesogenic diol in solution. The diol can be partially replaced by a polyether diol (flexible spacer). The polyaddition is carried out at an NCO/OH ratio of 1:1. Highly polar solvents (dimethylformamide, dimethyl sulfoxide, chloroform or dioxane) are used [29,30]. Anhydrous reagents should be applied. A two-step polyaddition process is preferred [20].

The first step of the polyaddition to obtain a prepolymer with terminal isocyanate groups is performed in solution or in melt [30]. In the two-step processes, 2,4-tolylene diisocyanate is commonly used.

The product is isolated from the solution by pouring into water [20] or precipitation with methanol [29]. Evaporation of the solvent to obtain cast films was also described.

DESCRIPTION OF THE SYNTHESIS [20]

2,4-Tolylene diisocyanate (2.16 mmol) was placed into a 50 mL flask equipped with a magnetic stirrer and a reflux condenser with drying tube. Poly(oxytetramethylene)diol M = 250 (0.54 mmol) was added as well as 3.2 ml dimethylformamide to obtain a 25% solution. The flask was placed in an oil bath at 70–80°C and the contents stirred for 1 h. Then 4,4′-bis(2-hydroxyethyoxy)biphenyl (1.62 mmol) was added and stirring was continued for 2 h at the same temperature. The solution was poured into 100 mL distilled water. The precipitated product was filtered off, washed three times with 50 mL methanol and dried at 100°C in vacuo for 4 h. Yield was 88% and M_n = 19,700.

Liquid crystalline polyurethanes with mesogenic side groups were synthesized in dimethylformamide solution in a one-step polyaddition process from equimolar amounts of a diisocyanate and a diethanolamine derivative with a mesogenic substituent [14].

Several syntheses of mesogenic reagents and of liquid crystalline polyurethanes were recently collected in an article [24].

Melt polycondensation was used to prepare liquid crystalline polyurethanes and copoly(urethane-carbonate)s [26,27]. An equimolar mixture of a mesogenic diol, 4,4′-bis(6-hydroxyhexoxy)biphenyl and 1,6-hexylene-bis (phenyl carbamate) was stirred at 180–190°C for 1 h under nitrogen, then at 190–200°C for 30 min at a pressure of 12–13 mm Hg and finally at 200–205°C for 30 min at 2 mm Hg. Phenol was distilled off. The polyurethane was purified by dissolving it in hot 1,4-dioxane, precipitating with water, filtrating, washing with methanol and drying at 50–60°C in vacuo.

To obtain copoly(urethane-carbonates), the bis-carbamate was partially replaced by the corresponding bis-carbonate.

PROPERTIES

Investigations of properties of liquid crystalline polyurethanes involve spectrometric methods (IR spectrophotometry in particular) as well as the molecular weight distribution measurements which do not differ from the chemical composition studies being performed in the case of conventional polyurethanes.

Moreover, there are methods of investigation which are specific for liquid crystalline polyurethanes. The most important one is the determination of transition temperatures (melting and isotropic transition). For that purpose, differential scanning calorimetry (DSC) and polarization microscopy is used. The DSC measurements give not only the transition temperatures, but also the enthalpy values of the phase transitions.

In addition to the thermal transitions, the morphology of liquid crystalline polyurethanes is investigated. The morphology is investigated by (1) polarizing microscopy equipped with a hot stage, (2) the SALS methods and (3) the WAXS method using an X-ray diffractometer.

In order to investigate the sorption behavior of liquid crystalline polyurethanes with hydrocarbons, sorption constants, diffusion coefficients and selectivity factors are determined using polyurethane films.

Structure-properties relationships for a series of liquid crystalline polyurethanes were studied [20,28,36–38]. The polyurethanes were synthesized from 4,4′-bis(2-hydroxyethyl)biphenyl BHBP (the mesogenic component), poly(oxytetramethylene)diol PTMO (flexible spacer) and 2,4-tolylene diisocyanate (2,4-TDI).

The effects of the method of synthesis (one-step or two-step) and the content of mesogenic agent on the properties of liquid crystalline polyurethanes were investigated [20]. Figure 1 presents the relationship between the composition of the polyurethanes and the phase transition temperatures (melting and isotropization temperatures) for one- and two-step methods. The DSC heating curves of the polyurethanes are shown in Figure 2.

Polyurethanes prepared by the one-step method usually exhibit higher T_m and T_i values than the polymers obtained by the two-step polyaddition. Moreover, the higher the PTMO content the lower are the phase transition temperatures for the obtained liquid crystalline polyurethanes, independently from the polyaddition methods applied.

The diffraction patterns of all investigated polyurethanes differ slightly from each other (Figure 3). The broad background came from an amorphous content and two weak diffraction lines from the crystalline part of the sample. The peaks at $2\theta = 18.6$ and 21.6 indicate crystalline behavior. Comparing the diffraction patterns of individual polyurethanes, it can be seen that the highest intensity of deflected lines on crystalline planes occurs with the LCP 1.00 (1) polymer con-

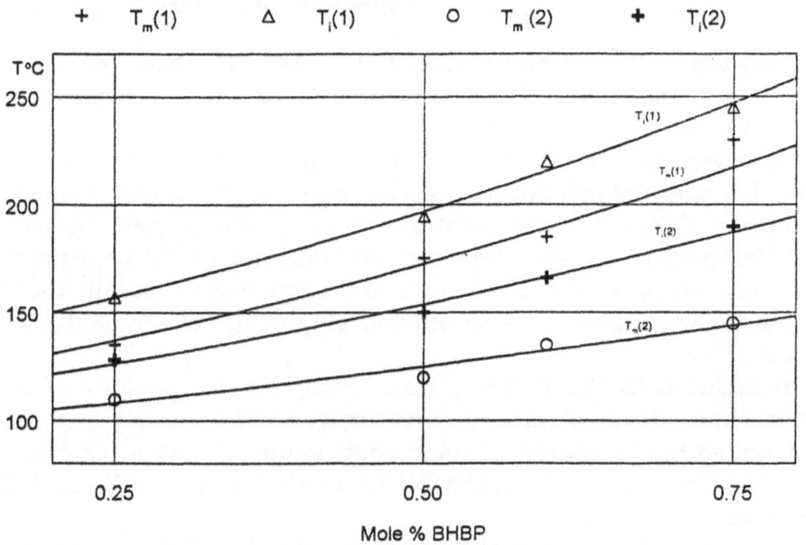

Figure 1. Dependence of melting point T_m and isotropic transition temperature T_i on the content of BHBP and on the method of synthesis; (1) one step method; (2) two step method.

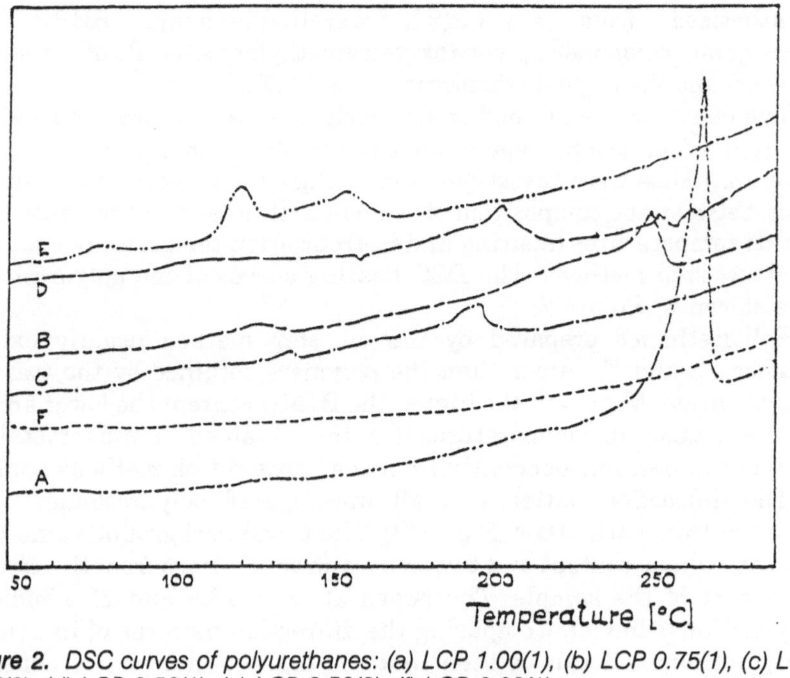

Figure 2. DSC curves of polyurethanes: (a) LCP 1.00(1), (b) LCP 0.75(1), (c) LCP 0.75(2), (d) LCP 0.50(1), (e) LCP 0.50(2), (f) LCP 0.00(1).

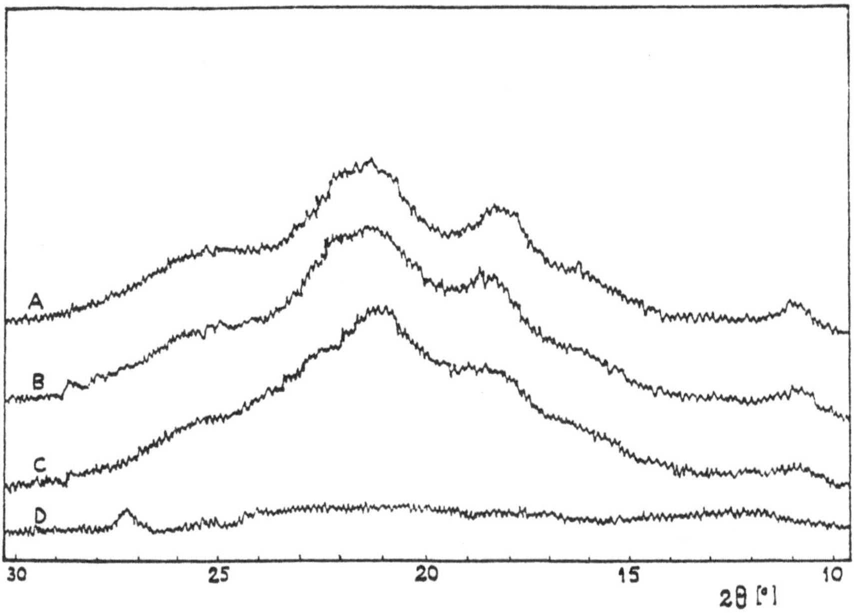

Figure 3. WAXS powder patterns: (a) LCP 1.00 (1), (b) LCP 0.75 (1), (c) LCP 0.75 (2), (d) LCP 0.00 (1).

taining no PTMO. The higher intensity of these lines was found if the polyurethanes were obtained by the one-step method. Increasing the PTMO content in the starting mixture of the hydroxyl comonomers resulted in a decrease in the intensity. For the LCP 0.40 (1) sample with 60% PTMO content, a very low intensity was found. The conclusion can be drawn that the crystalline properties of these polyurethanes are due to the occurrence of the BHBP–2,4-TDI sequences as well as for polyurethanes obtained by the one-step method. The one-step method favors the formation of longer BHBP-2,4-TDI sequences whereas the two-step method produces molecules with a more regular arrangement of the BHBP-2,4-TDI and PTMO-2,4-TDI sequences.

Investigations of the morphology of the liquid crystalline polyurethanes show that two of them [LCP 1.00 (1) and LCP 0.75 (2)] formed a spherulitic structure in the mesophase state. The scattering patterns show the presence of four symmetrical reflexes which are typical for spherulites. The mean radii of spherulites amount to 4.0 and 2.8 μm [LCP 1.00 (1) and LCP 0.75 (2), respectively]. They were calculated from the scattering patterns.

The effect of the flexible spacer length on some properties of liquid crystalline polyurethanes was investigated for a series of polyurethanes which differ in the flexible spacer length and BHBP content [28]. Poly(oxytetramethylene)diols of different molecular weights (PTMO, M_n = 250, 650, 1000, 2000) were used as flexible spacers. Figures 4 and 5 present the dependence of melting and isotropic transition temperatures on the length of the flexible spacer and on the content of BHBP, respectively. The values of the temperature at the intersection of the straight lines are the experimental data. The isotropization temperatures for all polyurethanes containing 75 mol% BHBP are similar, independently from the flexible spacer length. A decrease in the amount of the mesogenic agent results in a reduction of the melting temperature as well as the isotropization temperature. Increasing the molecular weight of the flexible spacer at a constant content of the mesogenic agent exerts the same effect. Both melting and isotropization temperatures are less sensible to changes of the PTMO spacer length at higher BHBP content.

The influence of molecular weight on some properties of liquid crystalline polyurethanes was studied for two series with the same content of BHBP and different lengths of flexible spacer (PTMO M_n = 250 and 2000) [36].

Polyurethanes with different molecular weight values (Low, Medium,

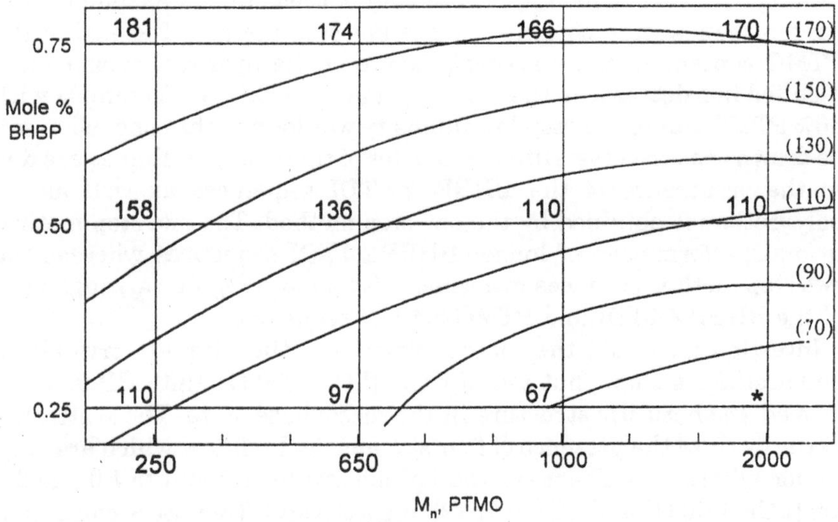

Figure 4. *Dependence of the melting point of the polyurethanes on the flexible spacer length and the BHBP content; *-no liquid crystalline properties.*

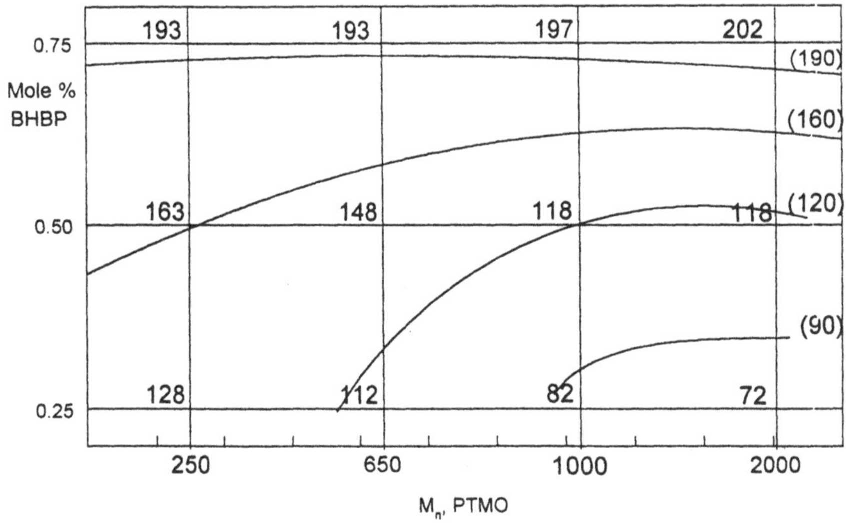

Figure 5. Dependence of the isotropic transition temperature on the flexible spacer length and the BHBP content.

High and Extra High) were synthesized by changing the isocyanate/diols ratio and the reaction time in the second step.

The thermal transition temperatures of the polyurethanes determined by DSC were correlated with the polarizing microscope observations. The values of transition temperatures and enthalpy changes of transitions are summarized in Table 3.

All the polyurethanes with the flexible spacer PTMO 250 and the polyurethanes LCP 0.75/2000/L and LCP 0.75/2000/M showed the transition in the solid state.

The influence of the molecular weight of the polyurethanes with different flexible spacer lengths on the melting and isotropization temperature is presented in Figure 6.

Generally, the higher the molecular weight, the higher are both T_m and T_i values. The values of these temperatures are more dependent on the molecular weight of the liquid crystalline polyurethanes than on the length of the flexible spacer.

The diffraction patterns of the polyurethanes show that the polyurethane LCP 0.75/250/L with the lowest molecular weight and flexible spacer PTMO 250 exhibit the highest crystallinity at room temperature. The peaks at 2θ = 18.6, 21.6, 23.0 and 24.3 indicate crystalline behavior (Figure 7).

Table 3. Thermal properties of the polyurethanes [36].

	Microscopic Observation		DSC Analysis						
Polymer	T_m (°C)	T_i (°C)	T_c (°C)	ΔH_c (J/g)	T_m (°C)	ΔH_m (J/g)	T_i (°C)	ΔH_i (J/g)	ΔT (°C)
LCP 0.75/250/L	143	163	123.0	9.21	144.8	1.30	165.8	1.39	20
LCP 0.75/250/M	162	181	107.4	4.10	—[b]		187.1	1.10[c]	19
LCP 0.75/250/H	181	193	111.9	3.50	—[b]		195.8	7.22[c]	12
LCP 0.75/250/X	198	210	123.6	4.93	—[b]		211.2	2.72[c]	12
LCP 0.75/2000/L	119	175	87.6	0.15	118.7	3.89	177.9	2.98	56
LCP 0.75/2000/M	174	188	119.7	0.36	171.5	0.29	188.2	3.06	14
LCP 0.75/2000/H	189	210	—	—	—	—	201.1	1.12	21
LCP 0.75/2000/X	191	219	—	—	—	—	220.5	0.93	28

T_c = transition temperature in the solid state. T_m = transition temperature from the solid phase to the mesomorphic phase. T_i = transition temperature from the mesomorphic phase to the isotropic liquid. ΔH_c, ΔH_m, ΔH_i = enthalpy changes of the above mentioned transitions. $\Delta T = T_i - T_m$ = temperature ranges of mesophases.
[b]The endotherms of melting and isotropization gave one very broad peak and the values for the melting temperatures were nonidentifiable in the DSC thermograms.
[c]Total value for the transitions from the solid state to the mesomorphic phase and to isotropic liquid.

For the polyurethane 0.75/250/M with medium molecular weight and flexible spacer PTMO 250, the peaks are weaker and the crystallinity of this polyurethane is lower. For the polyurethane LCP 0.75/250/X with the highest molecular weight, the intensity is very weak in comparison to these two polyurethanes. The diffraction pattern did not show any peaks. The diffraction patterns for the polyurethane with the flexible spacer PTMO 2000 showed similar dependences (Figure 8). Generally, for the polyurethanes with the flexible spacer PTMO 2000 the peaks were weaker than these for their counterparts with the flexible spacer PTMO 250.

The annealing effect on the thermal properties of the polyurethanes with different molecular weight, the same BHBP content and different flexible spacer length was investigated [37]. Figures 9 and 10 show the influence of the annealing time and temperature on the glass transition temperature, melting temperature and isotropization temperature of the polyurethane LCP 0.75/250/X (Mw = 61,950) with a BHBP content = 75 mol%, and the PTMO spacer Mn = 250.

Higher annealing temperatures as well as longer annealing times result in a considerable increase in the melting temperature and a weak increase in the isotropization temperature. This causes a decrease in the range of mesophase appearance.

Some of the LC polyurethanes varying in flexible spacer length and

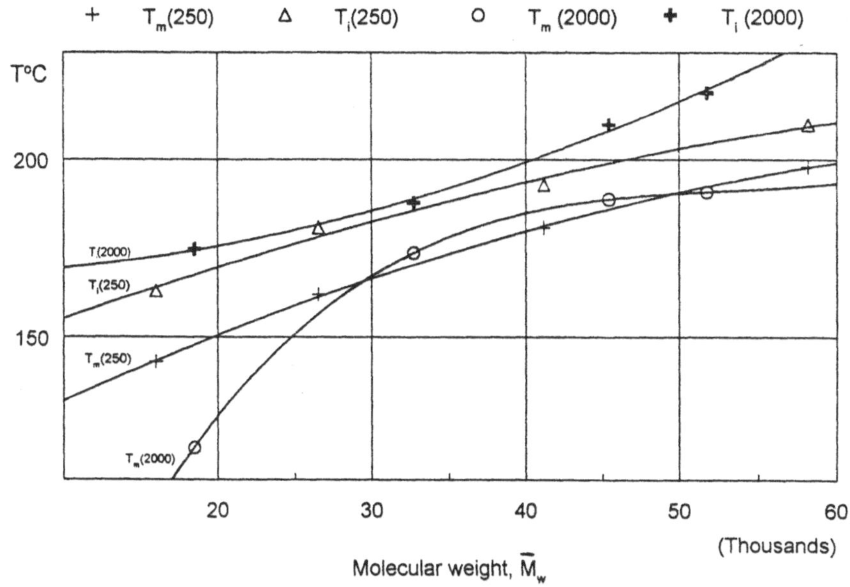

Figure 6. *Influence of average molecular weight on the melting temperature and isotropization temperature of polyurethanes with different spacer length.*

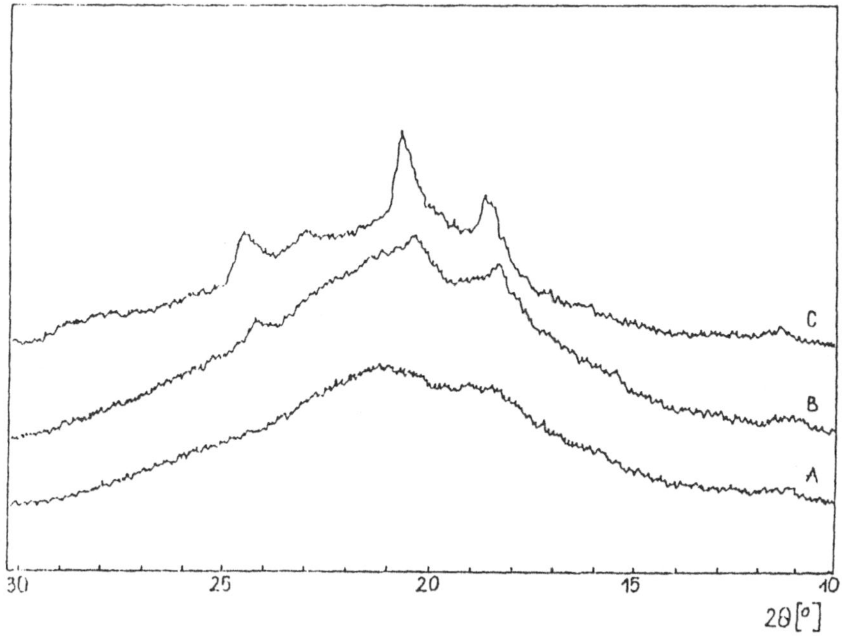

Figure 7. *WAXS powder patterns: (a) LCP 0.75/250/H, (b) LCP 0.75/250/M, (c) LCP 0.75/250/L.*

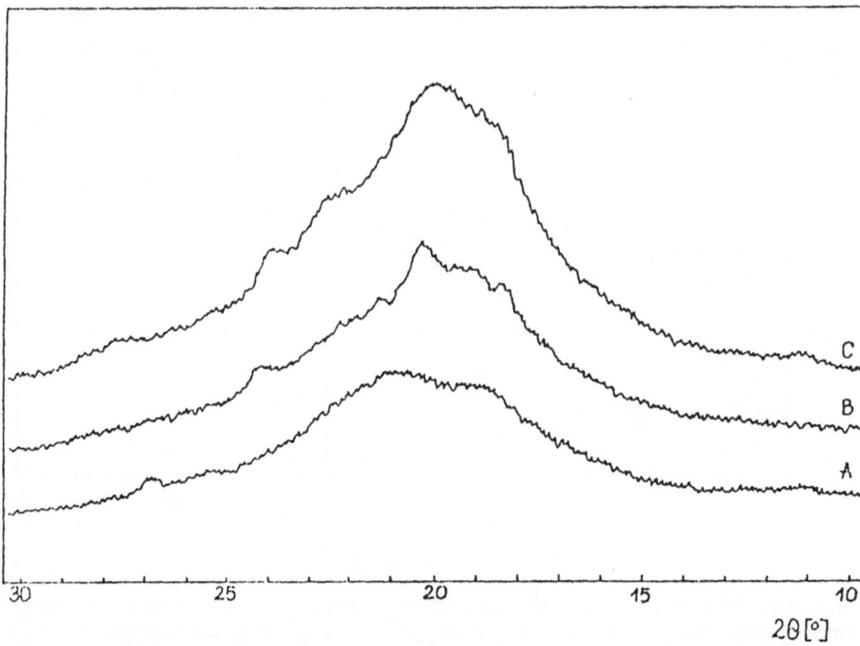

Figure 8. *WAXS powder patterns: (a) LCP 0.75/2000/H, (b) LCP 0.75/2000/M, (c) LCP 0.75/2000/L.*

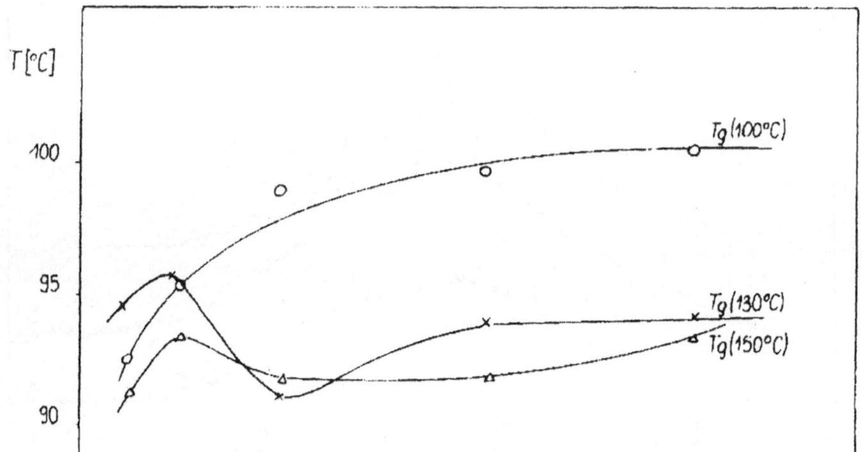

Figure 9. *Influence of the conditions of annealing on the glass transition temperature of polyurethane LCP 0.75/250/X.*

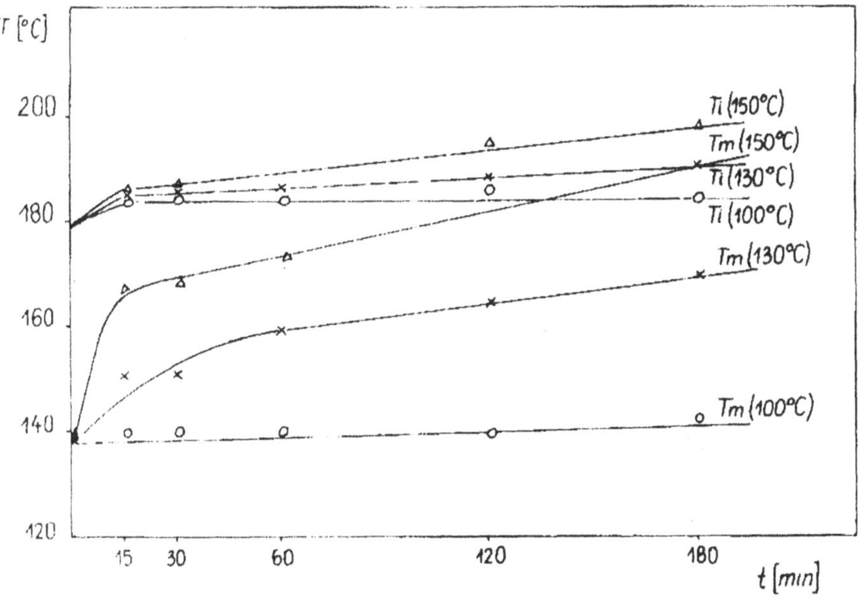

Figure 10. Influence of the conditions of annealing on the melting and isotropization temperature of polyurethane LCP 0.75/250/X.

mesogenic group content were selected and their transport properties with respect to the molecular and supermolecular structure were studied [38]. These polymers are targeted at the separation of aromatic hydrocarbons from their mixtures with aliphatic ones by means of the pervaporation method. The characteristics of the LC polyurethanes used in the transportation studies are given in Table 4.

It was found that structural differences among the particular LC polyurethanes which arose from different soft segment molecular weights and/or different initial molar ratios of the BHBP and PTMO diols used are strongly related to the appropriate differences in their sorption and diffusion coefficients. These differences are visualized by the sorption curves displaying the dependence of solvent sorbed vs. square root of time presented in Figures 11–14.

The transport results for various hydrocarbons and LC polyurethane materials are summarized in Tables 5 and 6.

It appears from the obtained data that diffusion and sorption of the aromatic and aliphatic hydrocarbons into the LC polyurethanes depends on polymer morphology as well as on the nature of penetrant. Although sorption and diffusion are confined to the soft domain, the impermeable hard domains influence overall transport by suppressing

Table 4. Characteristics of liquid crystalline polyurethanes used in transport studies [38].

Sample Code	Composition	Molecular Weight* M_n	Soft Segment (SS) Weight Fraction (%)	T_g** (°C)	Phase Separation Degree (%)†
LCP 0.75/ 2000	BHBP/PTMO 3:1 PTMO 2000	32,850	56.8	−85	100
LCP 0.50/ 2000	BHBP/PTMO 1:1 PTMO 2000	37,300	76.0	−85	100
LCP 0.75/ 1000	BHBP/PTMO 3:1 PTMO 1000	35,635	39.7	+10	49
LCP 0.50/ 650	BHBP/PTMO 1:1 PTMO 650	37,200	51.0	−32	71

*GPC, based on polystyrene standards.
**Glass transition temperature of the soft segments, DSC.
†Calculated as $T_{gHS} - T_g/T_{gHS} - T_{gSS}$ where $T_{gSS} = -85$°C (T_g of neat soft segments, experimental) $T_{gHS} = 100$°C (T_g of neat hard segments, experimental).

membrane swelling and increasing tortuosity of a diffusion path. It can be seen from the reduced values of S and D for LCP 0.75/2000 when compared with the respective values for LCP 0.50/2000, though the fairly pure soft segment matrix consisting mainly of the PTMO blocks is assumed to be present in both polyurethanes.

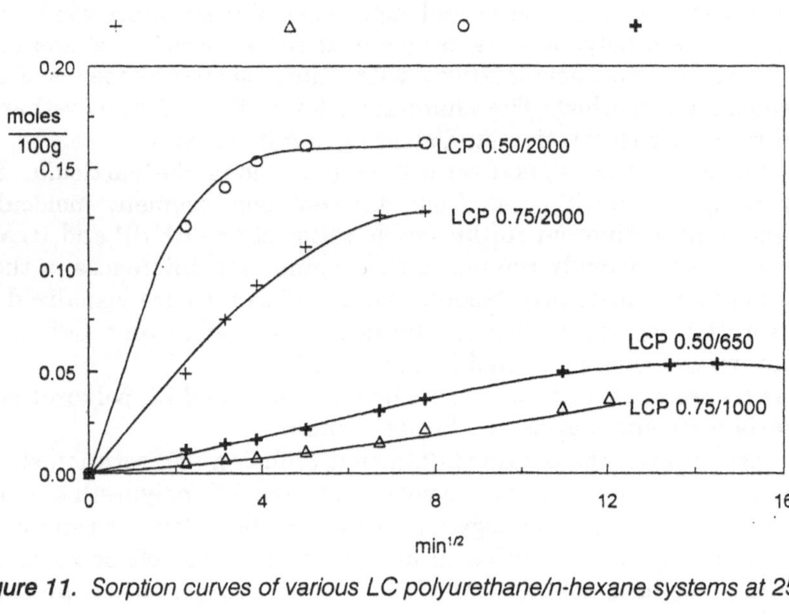

Figure 11. Sorption curves of various LC polyurethane/n-hexane systems at 25°C.

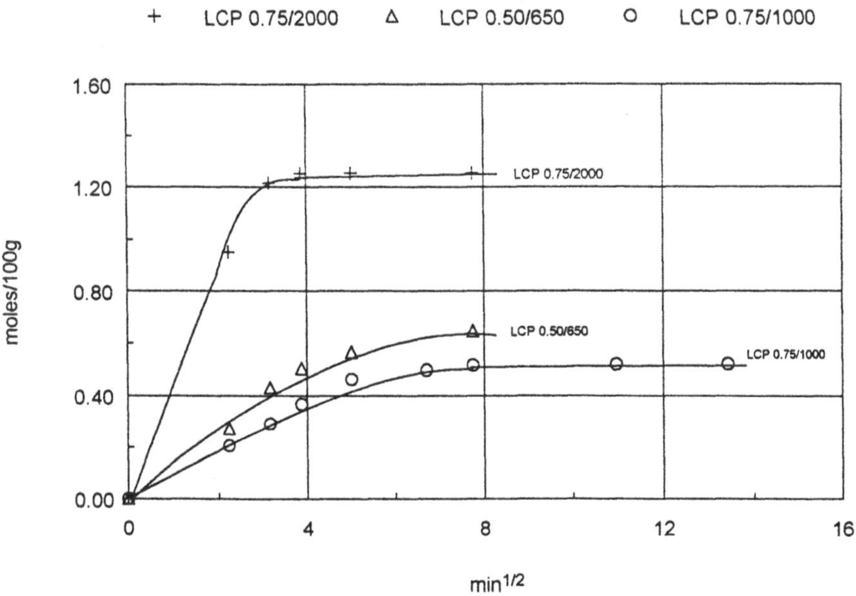

Figure 12. Sorption curves of various LC polyurethane/benzene systems at 25°C.

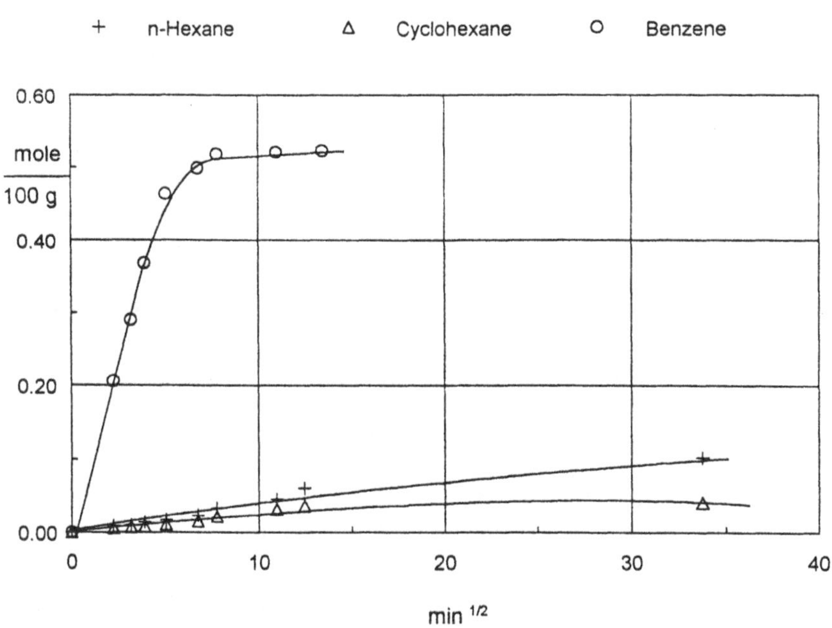

Figure 13. Sorption curves of various solvent /LCP 0.75/10000 systems at 25°C.

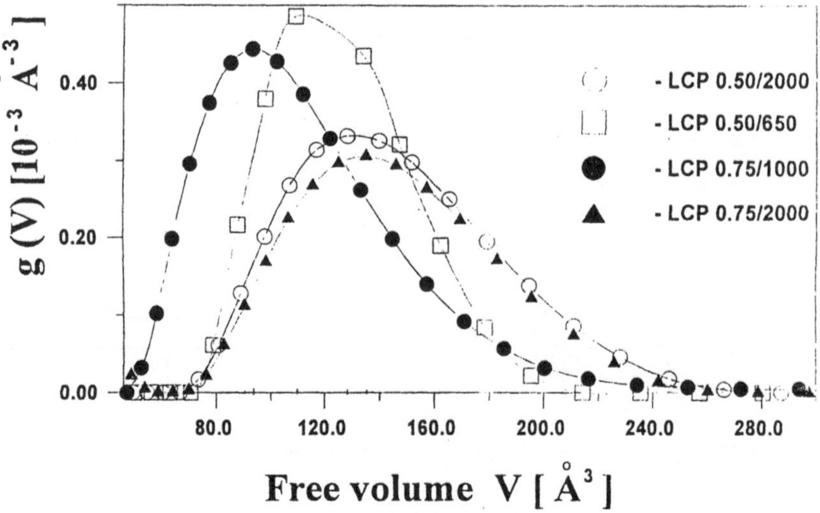

Figure 14. Free volume distributions for the studied LC polyurethanes.

On the other hand, correlations between the degree of phase mixing of the LC polyurethanes and the diffusion and sorption results were found to exist which can be attributed to the presence of highly polar mesogenic blocks in the soft domains. The existence of these blocks alters the chemical nature of the soft domains and thereby influences solvent sorptivity. It can also enlarge intermolecular interactions that makes the chains closer to each other and reduces the free volume thus decreasing solvent diffusivity. This latter suggestion is confirmed by the results of the investigations of LC polyurethanes by positron annihilation [39]. It appears from the curves shown in Figure 14 that the average size of free volume for the LC polyurethanes which seem to be

Table 5. Sorption constants (S) of LC polyurethanes at 25°C [38].

	Benzene		Cyclohexane		*n*-Hexane	
Sample Code	S (mol%)	S calcd. for 100% SS (mol%)	S (mol%)	S calcd. for 100% SS (mol%)	S (mol%)	S calcd. for 100% SS (mol%)
LCP 0.75/2000	1.21	2.13	0.29	0.51	0.14	0.25
LCP 0.50/2000	2.40	3.16	0.47	0.62	0.21	0.28
LCP 0.75/1000	0.54	1.36	0.11	0.28	0.05	0.13
LCP 0.50/650	0.58	1.14	0.13	0.26	0.05	0.10

Table 6. Diffusion coefficients and selectivity factors of LC polyurethanes at 25°C [38].

	Diffusion coefficient D · 10^7 (cm²/s)			Selectivity Factor α*	
Sample Code	Benzene	Cyclohexane	*n*-Hexane	Benzene/ Cyclohexane	Benzene/ *n*-Hexane
LCP 0.75/2000	3.00	0.73	1.12	17	23
LCP 0.50/2000	—	1.79	3.15	—	—
LCP 0.75/1000	1.33	0.06	0.11	109	131
LCP 0.50/650	2.02	0.09	0.25	100	94

*Calculated as D_AS_A/D_BS_B.

completely phase separated (LCP 0.75/2000 and LCP 0.50/2000) is higher than for the LC polyurethanes where phase mixing occurs (LCP 0.75/1000 and LCP 0.50/650).

The results of the analysis of the positron annihilation life time measurements given in Table 7 may also suggest the existence of two structurally distinct "soft" regions characterized by two different long-lived components in the lifetime spectra τ_3 and τ_4. Based on the values of the I_3 and I_4 intensities, which correspond to the positron annihilation in the region of the smaller and greater volume, respectively, it can be proved that the regions of mixed domains dominate in the LCP 0.75/1000 and LCP 0.50/650.

A general relationship was found to exist that the lower the value of the glass transition temperature T_g, the greater are the values of both average size of free volume V_{av} and the diffusion coefficient D. Moreover, the higher the degree of phase mixing, the higher are the differences in both sorption and diffusion coefficients between aromatic and non-aromatic solvents resulting in membrane materials with a high degree of permeation selectivity towards the investigated mixtures.

Table 7. Results of positron lifetime measurements [39].

	Lifetime [ns]				Intensity [%]			
LCP 0.75/2000	0.21	0.38	1.52	2.76	25.3	46.4	8.9	19.3
	(0.02)	(0.02)	(0.15)	(0.05)	(4.9)	(4.4)	(1.1)	(1.4)
LCP 0.5/2000	0.26	0.45	1.75	2.76	44.2	27.9	6.4	21.6
	(0.02)	(0.06)	(0.53)	(0.11)	(9.4)	(8.6)	(4.1)	(4.8)
LCP 0.75/1000	0.24	0.44	1.85	3.02	39.1	33.3	19.1	8.5
	(0.02)	(0.03)	(0.12)	(0.2)	(5.4)	(4.9)	(2.4)	(2.8)
LCP	0.27	0.58	2.30	5.93	55.1	19.3	25.1	0.5
	(0.01)	(0.04)	(0.04)	(1.65)	(2.2)	(2.0)	(0.3)	(0.3)

APPLICATIONS

The potential applications of liquid crystalline polyurethanes were mentioned in some articles dealing with synthesis, morphology and thermotropic properties thereof. As an example, elastomers and fibers in general [16] and melt processed fibers and films [33] can be mentioned. Moreover, fibers spun from solutions were described [11].

Liquid crystalline urethane ionomers, which were crosslinked with a melamine resin, could find applications as films and coatings [22].

Elastomeric films and materials with increased adhesion on metal surfaces were considered [30]. In the field of urethane elastomers, blends of liquid crystalline polyurethanes with common segmented polyurethanes should become an object of future applications (see previous chapter and Reference [31]).

In any proposed application, increased mechanical strength and modulus of elasticity is taken into account.

Membranes cast from solutions of liquid crystalline polyurethanes could find applications for organic liquids separations. It was expected that incorporation of mesogenic groups into a polyurethane chain should enhance the durability of the membrane and influence its transport properties [38]. Attention should be paid to the potential applications of amphiphilic liquid crystal polyurethanes [23].

The synthesis of high-molecular-weight liquid crystalline polyurethanes and copoly(urethane-carbonate)s through the transesterification of mesogenic diols with alkylene bis(phenylcarbamates) and their mixtures with the corresponding bis-carbonates may be considered as a future industrial process, leading to novel polymeric materials [25,26].

Potential applications include also electrooptical, piezoelectric and nonlinear optical devices.

Despite encouraging perspectives, no evidence could be found yet on an actual industrial application of liquid crystalline polyurethanes.

CONCLUSIONS

Polyurethanes with liquid crystalline properties can be synthesized from a variety of mesogenic isocyanates, diols and flexible spacers. An isocyanate-free method of synthesis can also be applied.

4,4′-Bis(2-hydroxyethoxy)biphenyl is a suitable mesogenic diol. Liquid crystalline polyurethanes with a broad range of phase transition temperatures can be obtained from that diol, 2,4-tolylene diisocyanate and poly(oxytetramethylene)diol as a flexible spacer.

The process of the synthesis (one-stage vs. two stage), content of the mesogenic diol, content and chain length of the flexible spacer, molecular weight of the polyurethanes as well as annealing of the polymer affect phase transition temperatures and morphology of the polymers.

The chemical composition of liquid crystalline polyurethanes has an influence on the liquid transport properties of membranes which could be suitable for the separation of organic liquids.

REFERENCES

1. 1991. *Liquid Crystallinity in Polymers.* A. Ciferri, ed., Raleigh, NC: North Carolina State University.
2. 1994. *Liquid Crystalline Polymers.* C. Carfagna, ed., Oxford, England: Elsevier Science Ltd.
3. 1988. *Liquid Crystalline Polymers.* N. A. Platé, ed., Moscow: Izd. Khimiya (in Russian).
4. Verbit, L. and G. A. Lorenzo. 1975. *Mol. Cryst. Liq. Cryst.,* 30:87.
5. Willner, L., F. Braun, M. Hess and R. Kosfeld. 1988. *Mol. Cryst. Liq. Cryst.,* 5:217.
6. Iimura, K., N. Koide, H. Tanabe and M. Takeda. 1981. *Makromol. Chem.,* 182:2569.
7. Kossmehl, G. and B. Hirsch. 1990. *Acta Polymer.* 41:597.
8. Tanaka, M., N. Yamagata, K. Kawaguchi and T. Nakaya. 1984. *Polym. Prepr. Jpn.,* 33:1647.
9. Sato, M. and S. W. Kantor. 1985. *Polym. Prepr. Jpn.,* 34:262.
10. Sato, M. 1985. *Polymer Digest, Jpn.,* p. 10.
11. Tanaka, M. and T. Nakaya. 1986. *Makromol. Chem.,* 187:2345.
12. Tanaka, M. and T. Nakaya. 1986. *Kobunshi Ronbunshu,* 43:311.
13. Tanaka, M. and T. Nakaya. 1987. *J. Macromol. Sci.-Chem. Phys.,* A–24:777.
14. Tanaka, M. and T. Nakaya. 1988. *Makromol. Chem.,* 189:771.
15. Onouchi, Y., Sh. Inagaki, H. Okamoto and J. Furukawa. 1990. *Kenkyu Hokoku–Asahi Garasu Zaidan,* 57:241; *Ch. Abs.* 116:21571 (1992).
16. Gerbi, D. J., C. W. Macosko, W. Mormann and S. Benadda. 1993. *Polym. Prepr. (Am. Chem. Soc., Div. Polym. Chem.),* 33(1):1109.
17. Kricheldorf, H. R. and J. Awe. 1988. *Makromol. Chem., Rapid Commun.,* 9:681.
18. Kricheldorf, H. R. and J. Awe. 1989. *Makromol. Chem.,* 190:2579.
19. Kricheldorf, H. R. and J. Awe. 1989. *Makromol. Chem.,* 190:2597.
20. Penczek, P., K. C. Frisch, B. Szczpaniak and E. Rudnik. 1993. *J. Polym. Sci., Polym. Chem. Ed.,* 31:1211.
21. Mormann, W. and S. Benadda. 1993. *Polym. Prepr. (Am. Chem. Soc., Div. Polym. Chem.),* 34(2):739.

22. Lorenz, R., M. Els, F. Haulena, A. Schmitz and O. Lorenz. 1990. *Angew. Makromol. Chem.*, 180:51.
23. Tanaka, M. and T. Nakaya. 1989. *Makromol. Chem.*, 190:3067.
24. Tanaka, M and T. Nakaya. 1993. *Adv. Urethane Sci. Technol.*, 12:1.
25. Tanaka, M. and T. Nakaya, in press.
25a. Haridas, K. R. and G. Radhakrishnan. 1995. *J. Polym. Sci., Polym. Chem. Ed.*, 33:901.
26. Sato, M., F. Komatsu, N. Takeno and K. Mukaida. 1991. *Makromol. Chem., Rapid Commun.*, 12:167.
27. Sato, M., T. Hirata, N. Takeno and K. Mukaida. 1991. *Makromol. Chem.*, 192:1139.
28. Szczepaniak, B., K. C. Frisch, P. Penczek, J. Mejsner, I. Leszczyńska and E. Rudnik. 1993. *J. Polym. Sci., Polym. Chem. Ed.*, 31:3223.
28a. Sun, Sh.-J., K. Y. Hsu and T. Ch. Chang. 1995. *J. Polym. Sci., Polym. Chem. Ed.*, 33:787.
29. Chiellini, E., G. Galli, S. Trusendi, S. A. Angeloni, M. Laus and O. Francescangeli. 1994. *Mol. Cryst. Liq. Cryst.* 243, 135.
30. Mix, R., H. Goering, G. Schulz, F. Grundemann and J. Gähde. in press.
31. Pohl, M.-M., R. Dany, R. Mix, J. Gähde and G. Hinrichsen. *Polymer,* in press.
32. Stenhouse, P. J., E. M. Valles, S. W. Kantor and W. J. MacKnight. 1989. *Macromolecules,* 22:1467.
33. Pollack, S. K., D. Y. Shen, S. L. Hsu, Q. Wang and H. D. Stidham. 1989. *Macromolecules,* 22:551.
34. Pollack, S. K., G. Smyth, F. Papadimitrakopoulos, P. J. Stenhouse and Sh.L. Hsu. 1992. *Macromolecules,* 25:2381.
35. Smyth, G., E. M. Valles, S. K. Pollack, J. Grebowicz, P. J. Stenhouse, S. L. Hsu and W. J. MacKnight. 1990. *Macromolecules,* 23:3389.
36. Szczepaniak, B., K. C. Frisch, P. Penczek, E. Rudnik and M. Cholińska. 1993. *J. Polym. Sci., Polym. Chem. Ed.*, 31:3231.
37. Szczepaniak, B., K. C. Frisch, P. Penczek and E. Rudnik. 1994. *J. Polym. Sci., Polym. Chem. Ed.*, 32:2559.
38. Wolińska-Grabczyk, A., I. Cabasso, B. Szczepaniak and P. Penczek. *J. Membr. Sci.*, submitted for publication.
39. Baranowski, A., M. Debowska, A. Jankowski, K. Jerie, P. Penczek, J. Rudiśka-Girulska, B. Szczepaniak, A Wolińska-Grabczyk, *Acta Phys. Polon.*, in press.

Synthesis and Structure-Property Relationships of Polyurethane Hydrogels

E. ***Haschke***

Busak + Shamban, Inc.
2531 Bremer Drive
Fort Wayne, IN 46803, U.S.A.

G. ***Hill***

Vistakon-Johnson & Johnson Vision Products, Inc.
5985 Richard Street
Jacksonville, FL 32216, U.S.A.

V. ***Sendijarevic****, S.* ***Wong****, and K. C.* ***Frisch***

Polymer Institute, College of Engineering and Science
University of Detroit Mercy
4001 West McNichols Road
Detroit, MI 48221, U.S.A.

ABSTRACT

Polyurethane hydrogels are utilized in a variety of applications including soft contact lenses, controlled release devices, semi-permeable membranes, and hydrophilic coatings. In order to prepare hydrophilic polyurethane materials of specific water absorption, light transparency, and mechanical properties, the formulation parameters of the polyurethane elastomers which control these properties must be understood. Polyurethane hydrogels based on polyether polyols, cycloaliphatic diisocyanates, and trifunctional, hydroxy-terminated crosslinkers were prepared using both solution and bulk polymerization methods. The effects of the chemical structure of the glycol, diisocyanate, crosslinker, and chain extender on the water absorption, mechanical properties, and optical clarity of the hydrogels were investigated. The crosslink density of the polyurethanes was varied by changing the molecular weight and concentration of the crosslinker, as well as the molecular weight of the poly(oxyethylene) and poly(oxypropylene) soft segments. The theoretical $\overline{M}_c$ values were calculated from the polyurethane formulation. The experimental values of the $\overline{M}_c$ were determined by compression modulus measurements of the polyurethane hydrogels. The interaction parameter, χ, and solubility parameter, δ, for the different structures of polyurethanes with water were determined using the experimental values of the $\overline{M}_c$. The experimental values for δ were confirmed by swelling of the hydrogels in formamide and in methanol.

INTRODUCTION

Hydrogel Types and Applications

HYDROGELS ARE POLYMERS CHARACTERIZED BY THEIR HYDROPHILICITY and insolubility in water. Hydrogels swell to an equilibrium state in water, but retain their shape due to the elastic restoring force of the polymer network.

Although the term "hydrogel" usually refers to a synthetic, water-swelling polymer, there also exist many natural hydrogels, such as those found in muscle, tendons, cartilage, intestines, and blood [1]. Natural hydrogels, such as cellulose, are used in pulp and paper production, artificial silk, membranes, and biomedical applications [2]. Hydrogels occupy a unique position in the field of biomaterials because of the ways in which the imbibed water strongly influences the transport, surface, and mechanical properties of polymers [3,4].

A familiar hydrogel is the dessert "JELLO," where the network is comprised of polymers derived from animal protein-gelatin [5]. The polymer network constitutes approximately 3 percent of the volume; the remainder is colored, flavored, sweetened water.

Hydrogels are utilized commercially as soft contact lenses, prosthetic devices, and function as drug-delivery matrices in controlled-release systems [6–11]. For application as a contact lens or other ocular device, the hydrogel must be transparent. The first hydrogel contact lenses were made from slightly crosslinked poly(2-hydroxyethylmethacrylate), which is still commonly used for soft contact lenses [12–14]. The contact lens must also be oxygen permeable. The rate of diffusion of dissolved oxygen through a soft contact lens is proportional to the concentration of water contained by the gel [15,16]. High oxygen permeability can also be achieved in soft contact lenses made entirely of silicone rubber, but at the expense of wettability [17,18]. The contact lens should also be resistant to protein absorption, which will result in clouding of the lens [19]. Silicone rubber, being inherently hydrophobic, interacts rapidly with lipids in the tear fluid, resulting in quick dewetting and potential adhesion to the cornea [20,21].

Hydrogels function as controlled release matrices two ways [7–11]. One method of controlled release is by the diffusion of the encapsulated material through the membrane, where the concentration gradient established by the difference in osmotic pressure on each side of the membrane controls the release. The other method is by the controlled solubility of the encapsulating membrane. As the membrane dissolves, the encapsulant is released. Both methods of release depend on the interaction parameter and solubility parameters of the encapsulant, the solvent, and the encapsulating polymer membrane.

There is substantial interest in synthetic hydrogels, of which various sorts are well known and have been described in the literature. In particular, homopolymers of acrylamide [22], *N*-vinylpyrrolidone [23], and hydroxyethylmethacrylate [24] have been studied. A number of investigators have produced hydrogel surfaces on conventional polymer substances by surface grafting [25,26]. The reaction product of the graft copolymerization of *N*-vinylpyrrolidone onto silicone rubber exhibited both hydrophilicity, high oxygen permeability, and improved wettability [27].

N-vinyl-2-pyrrolidone is readily polymerized by several means to produce a polymer, poly(*N*-vinyl-2-pyrrolidone), which is soluble in water and also in a small number of organic solvents [28]. Because of its hydrophilicity, *N*-vinyl-2-pyrrolidone is frequently copolymerized with hydrophobic monomers in the presence of a crosslinking agent to produce xerogels, which afford a wide variety of useful hydrogels when swollen in water [29–32]. Linear poly(*N*-vinyl-2-pyrrolidone) has several applications, in particular pharmaceutical ones [33]. However, although the crosslinked form of poly(*N*-vinyl-2-pyrrolidone) has a very high water absorbency, it is rarely used as a hydrogel because of its fragility [28].

Polyurethane Hydrogels

Polyurethane hydrogels are claimed to have applications in the biomedical area. Blair and Hudgin disclosed the use of hydrophilic polyurethanes as a soft contact lens material [15,34]. Gould and Johnston prepared interpenetrating polymer networks (IPNs) of polyurethanes and acrylates by polymerizing diacrylates in the presence of hydrophilic polyurethanes [35–40]. These systems also formed hydrogels and were claimed to have applications as contact lenses and surgical implants.

Polyurethanes have a broad applicability because their properties can be tailored by variation of their components, such as the polyols, diisocyanate, and chain extender, or crosslinker [41]. The biocompatibility, as well as blood compatibility, of polyurethanes has been well established as exemplified by the use of BIOMER, a linear polyurethane elastomer used to coat artificial hearts to prevent blood clotting [42]. Due to their physiological acceptability, good biocompatibility, stability over long implant periods, and excellent physical and mechanical properties, polyurethanes are widely used in medical applications [42–49]. Common features of the medically used polyurethanes are their water insolubility and their lack of reactive functional groups.

The incorporation of poly(ethylene glycol) (PEG) into a polyurethane results in increased hydrophilicity. PEG is a well known and industrially important polymer much used in the pharmaceutical and textile industries in its many commercially available high and low molecular weight grades [50–53]. The high solubility of PEG in water has led to its use in many other industries in the form of surface-active agents [54].

A detailed understanding of the interaction of water with this polymer would be expected since its introduction into industry occurred more than fifty years ago. This is not so, however. The compatibility of PEG with water is generally described in terms of rather vague hydrogen bonding or the Flory-Huggins interaction parameter, χ, both of which are inadequate in the context of PEG which has been found to form at least two specific hydrates [55,56]. One of the hydrates appears to be a distinct species which survives considerable aqueous dilution. A complete understanding of the compatibility or behavior of PEG solutions with water must take the formation of such specific hydrates into account.

Thermodynamic studies have shown that at high polymer concentration, the aqueous solution properties of PEG are controlled by negative entropies of dilution [52,57–59]. Such values indicate that a high degree of orientation of water molecules with respect to the polymer chain must exist. Results obtained by NMR [60,61] and IR [62,63], calorimetric [53,59,64,65] and Raman [62,63,66,67] studies have shown that in aqueous solution the PEG molecule forms a hydrogen bonded complex in which three water molecules are associated with each repeat unit. Evidence for a 1:1 complex has also been reported based on neutron scattering studies [68].

Liu and Parsons have studied the segmental environments of PEG in aqueous solution and suggested that each ethylene oxide unit in the PEG chain requires three molecules of water to form a hydrated complex [60,61]. Maxfield and Shepherd have also studied PEG in different states [66]. They proposed that a PEG molecule forms a hydrogen bonded complex in which three molecules of water are involved with each repeat unit and that this hydrogen bonded complex is destroyed if the solution is taken above the melting point of the PEG. Boehmke and Heusch carried out viscometric, calorimetric, and melting point studies of surface active agents containing short PEG chains and suggested the presence of strongly associated mono, tri, tetra, and hexa hydrates of the PEG unit [64]. A review of PEG interactions with water has been published with no mention of specific hydrate states [69]. Investigations of phase behavior and water binding in PEG solutions have been carried out using differential scanning calorimetry (DSC) [53,65] and light microscopy [70,71]. The estimated hydration number was found to be 2.7 for each PEG repeat unit.

Several authors have explained the anomalies of the heat of solution and dilution for PEG on the basis of its interaction with water [52,72–74]. An extensive study of the solution properties of high molecular weight PEG revealed that the polymer exhibits an inverse solubility-temperature relationship in aqueous media [33,51]. Bailey and Callard [57,58] correctly concluded that the negative entropy and heat of dilution of PEG in aqueous solution implies a high degree of polymer-solvent interaction and orientation of water molecules with respect to the polymer chain.

Graham et al. prepared crosslinked PEG hydrogels to study the degree of interaction of the polymer with a variety of solvents [75–77]. The PEG gels were formed by first reacting the PEG with 4,4′-diphenylmethane diisocyanate (MDI) and utilizing trimethylolpropane (TMP) to form urethane crosslinks. The polyurethane networks were subsequently allowed to swell in the solvents. The degree of swelling of the networks was plotted as a function of the solubility parameter, δ, of the solvent [75]. For nonhydrogen bonding solvents, e.g., hexane, chloroform, and toluene, and in aldehydes and ketones, the maximum swelling occurred at solubility parameter 9.8. In esters, the maximum swelling occurred at solubility parameter 11.0. In strongly hydrogen-bonding solvents, such as ethanol and benzyl alcohol, maximum swelling of the polymer occurred at 12.2. The plot of strongly hydrogen-bonding solvents contained three apparently anomalous points for methanol, formamide, and water, however, which could not be immediately explained.

Further investigation into the swelling of the polymer in different solvents at different temperatures revealed that the swelling solvents fit into two groups: those for which the degree of swelling increases with increasing temperature; and the anomolous solvents, water and formamide, whose swelling decreases significantly with temperature [56,75]. Graham et al. found that among the group of solvents which increase in their ability to swell the crosslinked PEG with temperature, there was in each case a sudden increase in the rate of change of swelling at about 27 °C [75]. Methanol, formamide, and acetic acid were found to show breaks in slope of their degree of swelling-temperature plots at temperatures between 20–55 °C. Water alone did not show such a break and displayed by far the steepest decrease in the degree of swelling as the temperature was increased. The normal behavior of gels with swelling solvents is to show an increase in the degree of swelling with temperature.

Graham et al. further showed that the degree of swelling of PEG gels in water is directly proportional to the negative reciprocal temperature [75]. Along with syneresis, or collapse of the gel with increasing temperature, Graham also found that the polymer still contained 112.6 pph of water at 100 °C, which he calculated corresponded to almost exactly three molecules of water per ethylene oxide unit [56,75].

Parameters That Determine Swelling

The degree of swelling of a hydrogel at equilibrium is governed by the chemical composition of the polymer and the crosslink density of the polymer network [2]. Hydrogels may be anionic, cationic, or nonionic in nature. Common to all hydrogels, however, is the presence of a hydrophilic moiety which allows the hydrogel to attract water. In general, the higher the concentration of hydrophilic sites within the polymer, the higher will be the swelling. The number of covalently bonded crosslinks per unit volume of the network is controlled by the concentration of polyfunctional polymer units. As the density of covalently bonded crosslinks within the polymer increases, the elasticity of the network decreases, as does the swelling.

The medium in which the hydrogels are swollen also influences their equilibrium degree of swelling. The temperature, pH, and ionic strength of the medium, as well as the external hydrostatic pressure, all contribute to the equilibrium concentration of water in the gel. The pH and ionic strength are more important in the case of charged gels [78].

From a phenomenological viewpoint, the equilibrium water content of a hydrogel is determined by the osmotic pressure, which is the driving force for the swelling. The osmotic pressure is the force exerted by the water per unit area of the hydrogel surface when the pure water is separated from the water imbibed by the gel before the establishment of equilibrium [79]. The resultant concentration gradient is determined by the strength of polymer-polymer interactions relative to the strength of polymer-water interactions. The relative strength of the polymer-polymer to polymer-solvent interactions can be quantitatively assessed by knowledge of the polymer-water interaction parameter, χ, and the solubility parameter, δ, for both the polymer and the solvent [80].

Acting in opposition to the osmotic pressure is the elastic restoring force of the polymer network which behaves as a spring causing the network to contract [81,82]. The sum of the forces acting on the gel can be represented by the following expression:

$$P_{\text{net}} = P_{\text{os}} - P_{\text{el}} \tag{1}$$

where P_{net} is the net pressure experienced by the gel, P_{os} is the osmotic pressure, and P_{el} is the elastic restoring pressure. At equilibrium, the sum of the forces acting on the gel is equal to zero, therefore:

$$P_{\text{os}} = P_{\text{el}} \tag{2}$$

The value of the pressure required to maintain equilibrium is referred to as the swelling pressure [83].

From a thermodynamic standpoint, swelling is concerned with the equilibrium between phases. In a normal solution, equilibrium is established between the pure phase, which is usually a solid, and the mixed phase, which consists of the solid solute dissolved in the solvent [84]. In the case of a hydrogel, the pure phase consists of pure water and the mixed phase consists of the hydrogel swollen with water. In order for swelling of the polymer to occur, the Gibb's free energy of mixing, ΔG_m, must be less than zero. The Gibb's free energy of mixing is determined by the sign and magnitude of the enthalpy of mixing, ΔH_m, and entropy of mixing, ΔS_m, at absolute temperature, T, according to [85]:

$$\Delta G_m = \Delta H_m - T\Delta S_m \tag{3}$$

For a hydrocarbon polymer such as natural rubber in a good solvent such as benzene, the swelling is accompanied by a large increase in configurational entropy due to the diffusing tendency of the polymer and solvent molecules. The enthalpy change for such a system is usually positive, i.e., the application of heat to the system will increase swelling, but is negligible in comparison with the positive entropy change [86]. For water-swelling substances, the entropy change is small and is often negative, due to the highly ordered structure of water, which prefers other water molecules as neighbors instead of diffusing with molecules of the polymer to produce a random mixture [87]. For this reason, water-swelling polymers sometimes become hazy after swelling due to an increased degree of crystallinity [6]. The driving force for swelling in water-swelling polymers is the exothermic heat of mixing, which results from the attractive hydrogen-bonding forces between molecules of the polymer and water. In both cases, swelling is accompanied by a negative change in the Gibb's free energy of mixing, but for thermodynamically different reasons.

The Gibb's free energy of mixing a polymer with a solvent can also be calculated from the Flory-Huggins equation, which expresses the free energy change accompanying mixing in terms of the volume fractions of the polymer and solvent as follows [88]:

$$\Delta G_m = RT\{\ln(1 - \phi_2) + \phi_2 + \chi\phi_2^2\} \tag{4}$$

where ϕ_1, which is equal to $1 - \phi_2$, is the volume fraction of solvent, ϕ_2 is the volume fraction of polymer, and χ is polymer-solvent interaction parameter which is characteristic of specific solvent-polymer pairs. The Flory-Huggins equation shows that if $\chi < 0.5$, ΔG_m is negative for all values of ϕ_2, i.e., the polymer is soluble. If $\chi > 0.5$, there exists a particular value of ϕ_2 for which $\Delta G_m = 0$; this represents the condition for limited or equilibrium swelling.

The Flory-Huggins relationship as it is presented above is applicable only when the polymer is not crosslinked. In the case of crosslinked polymers, which swell but do not dissolve when placed in a good solvent, the configurational entropy change of the network from swelling must also be taken into account. The modified Flory-Huggins relationship which considers configurational network entropy change is as follows [88,89]:

$$\Delta G_m = RT\{\ln(1 - \phi_2) + \phi_2 + \chi\phi_2^2 + (\varrho V_1/\overline{M}_c)(\phi_2^{1/3} - \phi_2/2)\} \tag{5}$$

where ϱ is the density of the unswollen polymer, V_1 is the molar volume of the solvent, and $\overline{M}_c$ is the molecular weight between crosslink points of the network. The condition for equilibrium swelling, $\Delta G_m = 0$, yields:

$$\chi = [-\ln(1 - \phi_2) - \phi_2 - (\varrho V_1/\overline{M}_c)(\phi_2^{1/3} - \phi_2/2)]/\phi_2^2 \tag{6}$$

The value of the χ parameter for a polymer-solvent pair reflects the compatibility of the polymer with the solvent. For a crosslinked polymer, the compatibility of the system increases as the χ parameter nears a value of 0.5, for which the degree of swelling of the polymer reaches a maximum.

A semi-empirical means of estimating the degree of compatibility of a polymer-solvent pair is by the determination of solubility parameters. Solubility parameters can be calculated using the Hildebrand-Scott equation [90]:

$$\Delta H_m = V\phi_1\phi_2[(E_1^v/V_1)^{1/2} - (E_2^v/V_2)^{1/2}]^2 \tag{7}$$

where V is the volume of the mixture of polymer and solvent, E_1^v is the molar heat of vaporization of the solvent, E_2^v is the molar heat of vaporization for the polymer, V_1 the molar volume of the solvent, and V_2 the molar volume of the polymer. Heats of vaporization for high molecular weight materials are not available and cannot be determined experimentally. The quantity E^v/V is referred to as the cohesive energy density (C.E.D.) of a substance and represents the energy required to overcome attractive forces between molecules of the substance. The solubility parameter, δ, of a substance is defined as $(E^v/V)^{1/2}$. The values of the solubility parameters for many common solvents have been determined from the isothermal vaporization of the saturated liquid to the ideal gas state at constant volume [80]. The solubility parameter of a polymer can be estimated from the degree of swelling of the polymer in a solvent of known solubility parameter. The solvent in which a polymer swells to its highest degree is assigned the solubility parameter of that solvent [91].

The solubility parameter, δ, is related to the χ parameter as follows [88,92]:

$$\chi = [V_1(\delta_1 - \delta_2)^2]/RT \tag{8}$$

Where V_1 is the molar volume of the solvent, δ_1 is the solubility parameter of the solvent, δ_2 is the solubility parameter of the polymer, R is the ideal gas constant, and T is the absolute temperature. Thus, by knowing the χ parameter of a particular polymer-solvent pair and the solubility parameter of the swelling liquid, the solubility parameter of the polymer can be calculated.

The χ parameter for a specific polymer-solvent pair can be calculated from Equation (6), if the $\overline{M}_c$ value of the polymer network is known. The $\overline{M}_c$ value can be determined experimentally by two methods [93]. One method is using equilibrium swelling data and Equation (6), for which the χ parameter must be known. The other method is by measurement of the equilibrium shear modulus of the polymer [28,94,95]. The shear modulus, G, of the polymer is related to the $\overline{M}_c$ of the polymer network:

$$G = \varrho RT/\overline{M}_c \tag{9}$$

where ϱ is the density of the unswollen polymer, R is the ideal gas constant, and T is the absolute temperature. The shear modulus of the polymer can be determined experimentally from a set of equilibrium force-extension values. To eliminate the contribution of intermolecular secondary forces to the modulus measurement, and to accelerate the time required for the displacement to reach an equilibrium value, the force-extension values are usually recorded with the polymer swollen to equilibrium in a good solvent. Under these conditions, the modulus value measured is due solely to covalent crosslinks between the polymer chains. The effective number of crosslinked units of the polymer, V_e, is related to the $\overline{M}_c$ through the network segment density, which is defined as V_e/V_0, where V_0 is the volume of the unswollen polymer as follows:

$$\overline{M}_c = \varrho V_0/V_e \tag{10}$$

The χ parameter and solubility parameter of the polymer network are representative of the polarity of the polymer and the degree of crosslinking. In the case of hydrogels, the solubility parameter reflects the ratio of hydrophilic/hydrophobic interactions and can be used as a tool to compare hydrogels of different compositions.

Present Work

A polyurethane hydrogel is an expanded, lightly crosslinked network of hard and soft segments surrounded by the swelling medium, water. The swelling of a polyurethane elastomer can be represented schematically as shown in Figure 1 [6]. The long, aligned wavy lines in the upper portion of the figure represent the crystallizable soft segment fraction of a polyurethane elastomer before swelling in water. Regions A and B in the figure represent the hard segment domains which are the crosslink points of the polymer network. After swelling with water, the previously aligned soft segment regions expand with the individual polymer chains becoming separated by molecules of water. The network is restricted from infinite swelling or dissolution through the crosslinking of the hard segments.

The work of Graham and others [6,8–11,56,77] demonstrated the hydrophilicity of PEG in polyurethane systems. In our previous work [96], the structure-property relationships of polyurethane hydrogel systems were investigated in regard to mechanical properties and

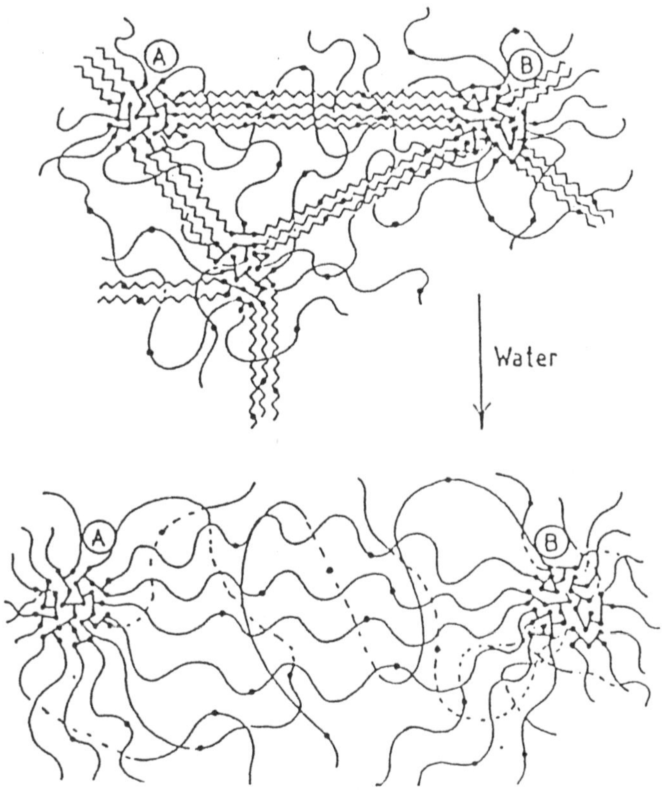

Figure 1. Schematic representation of the swelling of a polyurethane elastomer. (Reproduced with permission from Graham, N. B. and McNeill, M. E. Makromol. Chem., Macromol. Symp. *1988, 19, p. 258).*

swelling behavior of a large number of polyurethane systems. However, the formulation parameters which complement the hydrophilic nature of PEG were not investigated in a systematic effort to design polyurethane hydrogels which have specific water absorption, light transparency, and mechanical properties. The ability of polyurethanes to form hydrogels of predictable and reproducible water absorption, optimal clarity, and mechanical properties is critical to such applications as contact lenses and separatory membranes, where the polymer must absorb a specified quantity of water while maintaining its structural integrity.

The hydrophilic-hydrophobic balance within the polyurethane determines the solubility parameter of the polymer and its compatibility with solvents. In order to utilize hydrogels based on polyurethane as controlled release devices, the solubility parameters and polymer-solvent interaction parameters of the polyurethanes must be defined. Little information regarding the solubility and interaction parameters of the polyurethane hydrogels can be found in the chemical literature due to the wide variation of polyurethane structures and formulations.

In the present work, the polyurethane-water interaction parameters, χ, were determined by experimental measurement of the $\bar{M}_c$ of the polymer network utilizing compression modulus data. The solubility parameters, δ, of the polyurethanes were calculated based on the measured χ parameters. The assigned solubility parameters were confirmed by swelling the polyurethanes in solvents of known solubility parameters.

The ratio of bound (non-freezing) water to unbound (freezing) water was determined on a limited number of hydrogels using differential scanning calorimetry (DSC). The ratio of freezing/non-freezing water in polyurethane hydrogels as it relates to the polymer structure is of academic and practical significance.

EXPERIMENTAL

All materials utilized in the preparation of the polyurethane elastomers and hydrogels are listed in Table 1. The preparation of polyurethane elastomers and hydrogels, the determination of swelling, and measurement of stress-strain properties were described in detail in the previous paper [96]. The swelling ratios were calculated according to:

$$\text{Swelling ratio} = (W_f - W_i)/W_i \tag{11}$$

where W_f is the swollen weight of the hydrogel and W_i is the dry weight of the elastomer before swelling.

Determination of Polymer Density

The densities of the polyurethane elastomers were determined on dry 10 to 50 mg specimens. The density of the polymer was obtained from the specific gravity by multiplying the specific gravity value by the density of water, which is 0.99707 g/mL at 25 °C [97]. The specific gravity was measured according to ASTM D 297, Section 16. The water used for the determination was degassed by boiling for one hour.

Table 1. Materials.

Compound	Structure	Molecular Weight ($\bar{M}_n$)	Name	Supplier
PEG	$HO{-}(CH_2CH_2O)_n{-}H$	400, 600, 1000, 1500, 2000, and 4000	Poly(oxyethylene) glycol	BASF
PPG	$HO{-}(CH_2CH(CH_3)O)_n{-}H$	1000 and 2000	Poly(oxypropylene) glycol	BASF
TMP	$CH_3CH_2C{-}(CH_2OH)_3$	134	Trimethylolpropane	Eastman-Kodak
$H_{12}MDI$	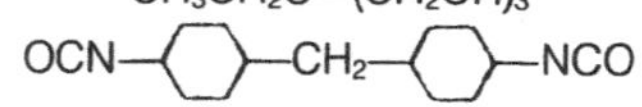	262	4,4′-Methylene-bis (cyclohexylisocyanate)	Miles
IPDI	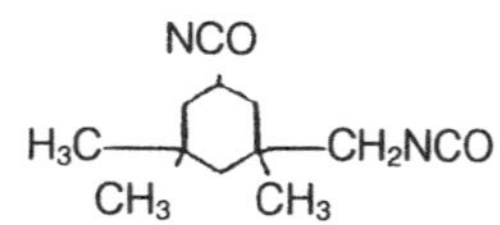	222	Isophorone diisocyanate	Huls
EG	$HOCH_2CH_2OH$	62	Ethylene glycol	Spectrum
P-726	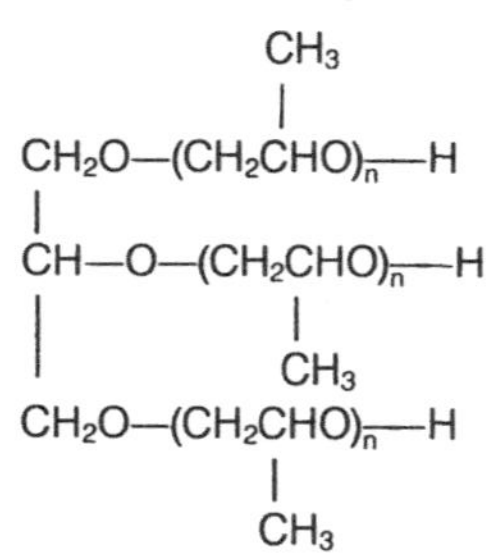	2900	Poly(oxypropylene) triol	BASF

(continued)

Table 1. (continued).

Compound	Structure	Molecular Weight ($\bar{M}_n$)	Name	Supplier
T-12	$(C_4H_9)_2Sn(O_2C(CH_2)_{10}CH_3)$	—	Dibutyltin dilaurate	Air Products
1,4 Dioxane	—	—	Dioxane	Baker
2-Ethoxyethyl acetate	—	—	Cellosolve acetate	Eastman-Kodak
Sodium chloride	$NaCl$	—	Sodium chloride	Baker
Boric acid	H_3B_3	—	Boric acid	Baker
Sodium borate	Na_3BO_3	—	Sodium borate	Baker
EDTA	$Na_2C_{10}H_{14}N_2O_8$	—	Ethylenediaminetetra-acetic acid, disodium salt, dihydrate	Baker

Determination of Polymer Crosslink Density

The crosslink density of the polyurethane elastomers, which can also be expressed as the molecular weight between crosslinks ($\bar{M}_c$), was determined by compression modulus measurements of elastomer specimens in the swollen state in water. The compression modulus of the hydrogels was determined using a thermomechanical analyzer (TMa, DuPont 2940) by measuring the displacement of a quartz probe placed on top of the specimen as a linear function of the applied force. The value of the displacement used for calculation of the modulus was the equilibrium value, obtained by allowing the specimen to equilibrate at each value of the force. The equilibration time required was found to be about twenty minutes.

The specimens used for the determination were cylindrically shaped hydrogels, cut using a standard die of 5.50 mm diameter. The diameter of the quartz probe was also 5.50 mm, allowing for complete contact between the probe and the sample. The thickness of the specimens varied between 0.5 to 2 mm. The hydrogel was placed in a 1.5 cm diameter cup filled with distilled water which was placed on the glass stage of the TMA beneath the quartz probe. The initial thickness of the specimen was accurately measured using the linear displacement measuring capability of the TMA after the application of a 25 g (0.25 N) external weight to the tray of the TMA, which insured good contact between the probe and the sample.

The modulus measurements were made by the application of a programmed force to the specimen by the TMA in increments of 0.1 N in the range of 0.1 to 0.5 N. The equilibrium displacement was recorded by the TMA in units of μm. The force applied to the specimen was controlled such that the compressive displacement did not exceed fifteen percent, in order that the stress-strain behavior of the elastomer remained linear. All measurements were conducted at 25 °C (298 K). The equilibrium shear modulus was calculated from the slope of the force-displacement curve as follows [28,94,95]:

$$G = [(F_2 - F_1)/(\pi d^2/4)]/[(L_2 - L_1)/L_1] \quad (12)$$

where G is the equilibrium value of the shear modulus, F_1 is the applied force at time 1, F_2 is the applied force at time 2, d is the diameter of the dry test specimen, L_1 is the thickness of the specimen at time 1 and L_2 is the thickness of the specimen at time 2.

The $\bar{M}_c$ value of the network was calculated from the equilibrium shear modulus values according to [88]:

$$\overline{M}_c = [(\varrho RT)/G]\phi_2^{1/3} \tag{13}$$

where ϱ is the density of the dry polymer, R is the ideal gas constant, 8.314 J K^{-1} mol^{-1}, T is the absolute temperature (K), and ϕ_2 is the volume fraction of polymer in the swollen system.

Determination of Freezing and Non-Freezing Water

The relative ratio of non-freezing (bound) water to freezing (unbound) water in the hydrogels swollen in deionized water was determined using differential scanning calorimetry (DSC, DuPont 910). The DSC was clarified so that the onset point of melting distilled, deionized water was exactly 0.00°C. Hydrogel specimens of 10 to 20 mg were patted dry with Kimwipes to remove surface water and placed in open aluminum sample pans. Heat flow was recorded relative to an empty sample pan. The hydrogels were analyzed in the temperature range of −60°C to +25°C using a heating rate of 5°C/min.

The relative percentages of freezing and non-freezing water in the hydrogels were calculated by the quantitative integration of the melting water endotherm from its onset point at −40°C to 0°C, and from 0°C to the endpoint of the endotherm at 20°C, and from the equilibrium water content of the hydrogels from gravimetric swelling measurements. The integration was performed by cutting and weighing paper copies of the DSC endotherms of an analytical balance. The percentage of non-freezing (bound) water was calculated from the difference between the total quantity of water contained by the gel and the quantity of water which melted above 0°C.

Calculation of Polyurethane-Water Interaction Parameters

The polymer-water interaction parameter, χ, was calculated from the equilibrium swelling data of the hydrogels in distilled water according to Equation (6) [88]. The $\overline{M}_c$ value used in the calculation is the theoretical value based on the polyurethane formulation [93].

The volume fraction of polymer, ϕ_2, in the hydrogel was calculated from the equilibrium swelling data according to [88,89]:

$$\phi_2 = (m_p/\varrho_p)/[(m_p/\varrho_p) + (m_w/\varrho_w)] \tag{14}$$

where m_p is the mass of the polymer in the hydrogel, ϱ_p is the density of the unswollen polymer, m_w is the mass of the water in the hydrogel, and ϱ_w is the density of water at 25°C (0.99707 g/mL) [97].

Calculation of Polyurethane-Water Solubility Parameters

The polymer-water solubility parameter, δ, was calculated from the χ parameter for the hydrogel using Equation (8). The solubility parameter for water is 23.4 $(cal/cm^3)^{1/2}$ [97].

Calculation of H_2O/EO Mole Ratios

The molar ratio of water/oxythylene groups was calculated from the equilibrium swelling data of the hydrogels in distilled water.

Calculation of Theoretical $\overline{M}_c$ Values

The theoretical value of $\overline{M}_c$ was calculated based on the polyurethane formulation [93]. $\overline{M}_c$ is the average molecular weight between crosslink points of the network. The molecular weight of the segment is equal to twice the number of equivalents of diisocyanate and polyol per equivalent of crosslinker.

RESULTS AND DISCUSSION

Effect of Oxyethylene Group Concentration on Swelling, Transparency, and Mechanical Properties of Polyurethane Hydrogels

PEG-Based Hydrogels

The equilibrium water content of the polyurethane hydrogels is a function of the oxyethylene concentration of the system and the crosslink density of the polymer network. The oxyethylene concentration of the system is proportional to the PEG soft segment concentration. The crosslink density of the polymer network is related to the molecular weight between crosslinks of the network, $\overline{M}_c$. The equilibrium water content of the hydrogels would be expected to increase with increasing concentration of PEG and increasing $\overline{M}_c$. Polyurethane hydrogels prepared from H_{12}MDI, PEG's, and trimethylolpropane (TMP) (2/1/1 equivalent ratio) clearly exhibit this relationship (Tables 2 and 3, Figure 2). As the PEG concentration of the system increased from 40 to 88 wt% by increasing the molecular weight of the PEG soft segment, the equilibrium water content of the hydrogels increased from 11 to 80%. Over the same period, the $\overline{M}_c$ of the polymer network increased from 493 g/mol for PEG 400 to 2293 g/mol for PEG 4000. The water/

Table 2. Effect of PEG molecular weight on the elongation, tensile strength and water absorption of polyurethane hydrogels prepared in bulk by compression molding.

Designation	1	2	3	4	5	6
Polyurethane composition (eq.)						
$H_{12}MDI$	2.0	2.0	2.0	2.0	2.0	2.0
PEG 400	1.0	—	—	—	—	—
PEG 600	—	1.0	—	—	—	—
PEG 1000	—	—	1.0	—	—	—
PEG 1500	—	—	—	1.0	—	—
PEG 2000	—	—	—	—	1.0	—
PEG 4000	—	—	—	—	—	1.0
TMP	1.0	1.0	1.0	1.0	1.0	1.0
Hydrogel properties[a]						
% Saline	11.2	24.4	45.3	66.3	68.9	80.2
% Elongation	—	100	35	71	20	—
Tensile strength (MPa)	—	3.6	5.3	2.8	1.9	2.4
Appearance (c = clear, h = hazy, o = opaque)	c	c	c	h	h	o
Physical properties of polyurethane						
Moles EO	9.10	13.6	22.8	34.0	45.4	91.0
Moles H_2O	7.10	21.8	74.2	232	322	1038
H_2O/EO mole ratio	0.780	1.60	3.26	6.79	7.08	11.4
Theoretical $\overline{M}_c$ (g/mol)	986	1186	1586	2086	2586	4586

[a]Data previously reported [96].

Table 3. Effect of PEG molecular weight on the elongation, tensile strength and water absorption of polyurethane hydrogels prepared by solution polymerization in 1,4-dioxane. Solid/solvent weight ratio = 60/40.[a]

Designation	1	2	3
Polyurethane composition (eq.)			
$H_{12}MDI$	2.0	2.0	2.0
PEG 600	1.0	—	—
PEG 1000	—	1.0	—
PEG 1500	—	—	1.0
TMP	1.0	1.0	1.0
Hydrogel properties			
% Saline	19.5	54.3	62.2
% Elongation	303	87	45
Tensile strength (MPa)	2.43	1.81	1.22
Appearance	hazy	hazy	clear

[a]Data previously reported [96].

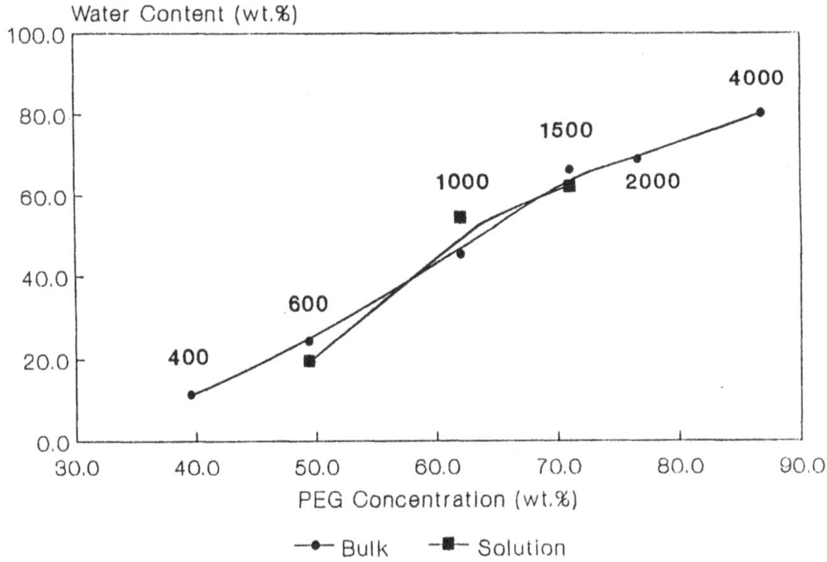

Figure 2. Effect of the PEG concentration on the water uptake of polyurethane hydrogels ($H_{12}MDI/PEG/TMP$ = 2/1/1 equivalent ratio) prepared in bulk and in solution using dioxane as solvent (40 weight percent solids). Bold-faced numbers on graph indicate PEG molecular weight.

oxyethelene mole ratio (H_2O/EO) increased from 0.78 to 11.4 by increasing the molecular weight of the PEG from 400 to 4000 (Table 2 and Figure 3). Based on these data, it can be concluded that the molecular weight between crosslinks ($\overline{M}_c$) has a great effect in determining the swelling of the hydrogels.

Elastomers prepared by both bulk and solution polymerization procedures exhibited similar water absorption abilities, indicating that the absorption of water is an inherent property of the material, characteristic of a specific formulation (PEG concentration), and is less dependent on the preparation method. The hydrogels which resulted from bulk polymerization were prepared by the prepolymer method, and those in solution were prepared by the one-shot method (Tables 2 and 3).

Large differences were exhibited in the elongation values between hydrogels prepared in bulk and those prepared in solution (Tables 2 and 3, Figure 4). The elongation of the hydrogels is less of an inherent property of the material and is strongly dependent on measurement conditions, sample dimensions, and other external variables. However,

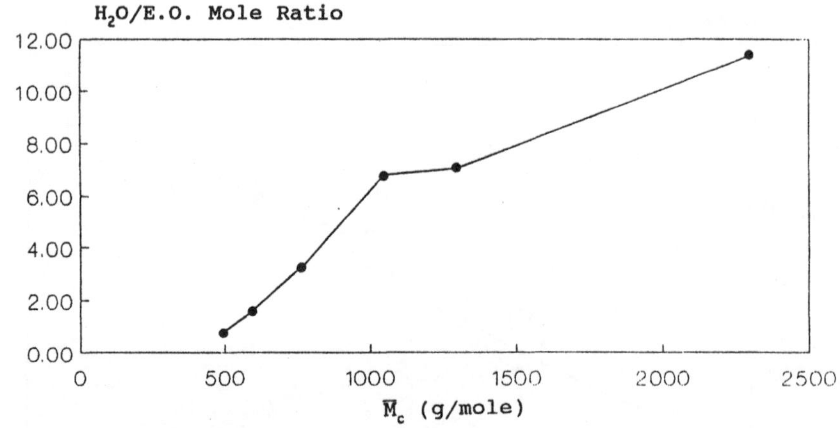

Figure 3. Effect of the molecular weight between crosslinks ($\overline{M}_c$) on the H_2O/oxyethylene mole ratio for polyurethane hydrogels ($H_{12}MDI/PEG/TMP = 2/1/1$) prepared in bulk by the prepolymer method.

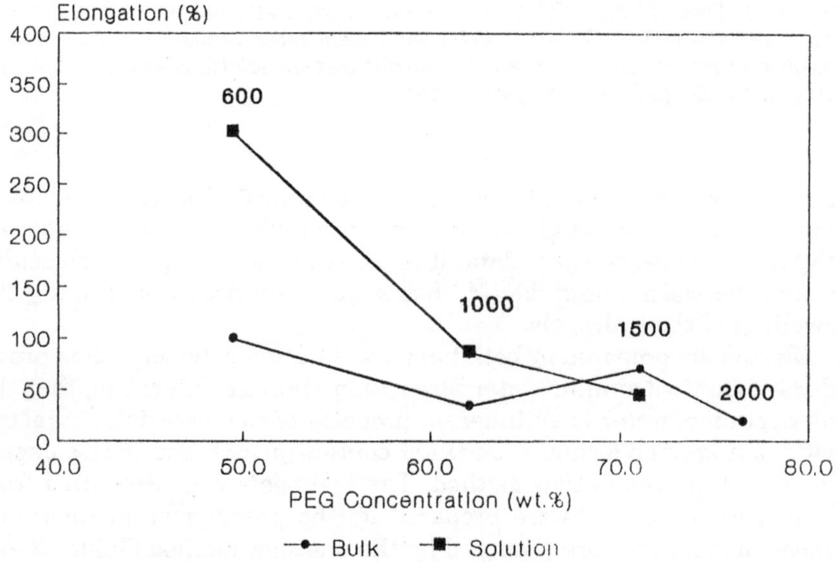

Figure 4. Effect of the PEG concentration on the elongation of polyurethane hydrogels ($H_{12}MDI/PEG/TMP = 2/1/1$ equivalent ratio) prepared in bulk and in solution using dioxane as solvent (40 weight percent solids). Bold-faced numbers on graph indicate PEG molecular weight.

hydrogels which were prepared by the prepolymer (bulk) method of polymerization should have a more regular network structure than hydrogels which were prepared by the one-shot (solution) polymerization. The difference in reactivity between the hydroxyl groups in the TMP molecule and the terminal hydroxyl groups of the PEG polyol would tend to create a more random crosslink pattern in the hydrogels prepared by the one-shot (solution) method. As a result of the less ordered structure, the hydrogels prepared in solution exhibited higher elongation. The elongation of the hydrogels decreased for samples prepared by both bulk and solution polymerization methods as the concentration of PEG in the system increased, and would be expected when the water content of the gel increases.

The tensile strength of the hydrogels exhibited behavior similar to the elongation values (Tables 2 and 3, Figure 5). Hydrogels which were prepared by the prepolymer (bulk) polymerization method exhibited higher tensile strength than those prepared by the one-shot (solution) polymerization method. Since the hydrogels prepared by the prepolymer polymerization method have a more ordered crosslink pattern, they should exhibit a higher tensile strength than hydrogels prepared

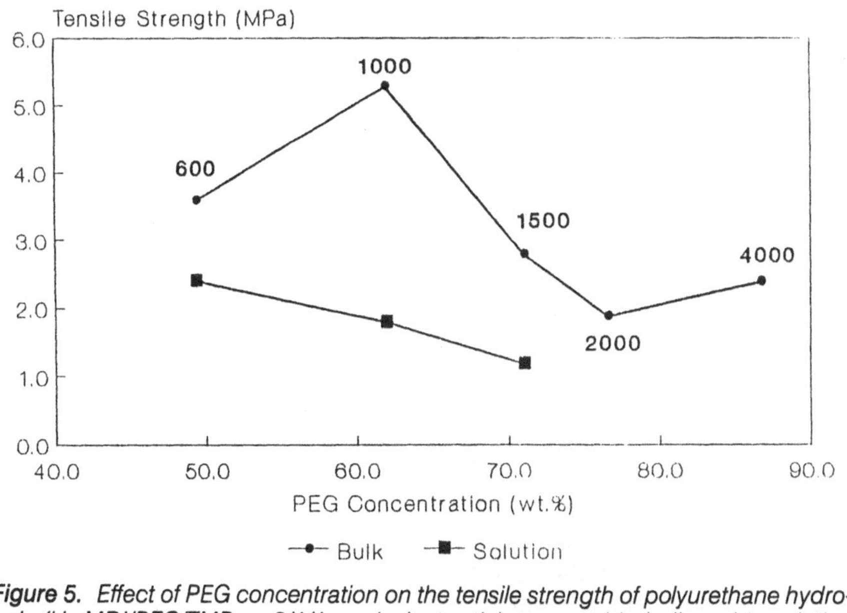

Figure 5. Effect of PEG concentration on the tensile strength of polyurethane hydrogels (H_{12}MDI/PEG/TMP = 2/1/1 equivalent ratio) prepared in bulk and in solution using dioxane as solvent (40 weight percent solids). Bold-faced numbers on graph indicate PEG molecular weight.

by the one-shot method. As the PEG concentration of the system increased, both the soft segment concentration and water content of the hydrogels increased, leading to decreased tensile strength. This relationship is generally seen to be true, although there are again differences between the bulk- and solution-prepared materials.

PEG and PPG-Based Hydrogels

Introduction of poly(oxypropylene) glycols (PPG) into the polyurethane network reduced the water content of the hydrogels; however, their transparency improved. Hydrogels prepared using only PEG were not transparent above PEG molecular weight 1000, due to the high degree of crystallinity of the soft segments [6]. PPG, being asymmetrical, disrupts the crystallinity and results in transparent hydrogels.

The elongation of the hydrogels increased for systems of both degrees of crosslinking as the PPG concentration of the systems increased, as would be expected as the water content of the gels decreased (Tables 4 and 5, Figure 6). The elongations of hydrogels prepared at the lower degree of crosslinking were higher than the elongations of hydrogels of the same composition which were more highly crosslinked. This trend would be expected due to an increased $\overline{M}_c$ value with decreased crosslinking, showing that the elasticity of the polymer network controls the mechanical properties of the hydrogels [81,98].

The tensile strength of the hydrogels of the more crosslinked system (2/1/1) was higher than the tensile strength of hydrogels which were less crosslinked (2.5/1.5/1), as expected [81,98]. The tensile strength of the hydrogels with the higher degree of crosslinking increased more sharply above 30 wt% PPG than hydrogels with the lower degree of crosslinking, as the PPG concentration of the system increased and the water content decreased (Tables 4 and 5, Figure 7).

Hydrogels prepared from H_{12}MDI, PEG 1500, PPG 2000, and TMP at equivalent ratios of 2/1/1 and 2.5/1.5/1 (H_{12}MDI/(PEG 1500 + PPG 2000/TMP) displayed a decrease in water absorption as the PPG concentration of the systems increased (Tables 4 and 5, Figure 8). Hydrogels based entirely on PPG soft segments did not absorb any measurable quantity of water. The PPG concentration of the systems was increased by substitution of PEG with PPG, which again shows that the water content of the hydrogels is proportional to the oxyethylene concentration of the system.

When substituting PEG 1500 with PPG 2000, however, the average number of repeat units–(CH_2CHX–O)–in the chain remains practi-

Table 4. Mechanical and physical properties of polyurethane hydrogels based on PEG 1500 and PPG 2000 [(PEG + PPG)/TMP/H_{12}MDI = 1/1/2 equivalent ratio] prepared by solution polymerization in 1,4-dioxane. Solid/solvent weight ratio = 60/40.

Designation	1	2	3	4	5	6
Polyurethane composition (eq.)						
H_{12}MDI	2.0	2.0	2.0	2.0	2.0	2.0
PEG 1500	1.0	0.9	0.8	0.75	0.67	0.47
PPG 2000	0.0	0.1	0.2	0.25	0.33	0.53
TMP	1.0	1.0	1.0	1.0	1.0	1.0
Hyrdogel properties[a]						
% H_2O	61.5	56.8	61.0	53.2	44.8	38.3
% Saline	62.2	60.6	57.2	54.4	41.9	37.8
% Elongation	45	65	74	114	203	231
Tensile strength (MPa)	1.22	0.82	0.88	1.08	3.14	3.60
Hardness (Shore A)	60	55	24	45	25	30
Appearance (c = clear)	c	c	c	c	c	c
% Non-freezing water	—	—	—	89.1	96.4	—
Physical properties of polyurethane						
Density (g/mL)	1.109	1.261	1.147	1.144	1.126	1.109
Moles EO	34.0	30.6	27.2	25.6	22.6	16.0
Moles H_2O	187	158	192	141	128	107
H_2O EO mole ratio	5.49	5.16	7.05	5.53	5.67	6.70
Theoretical $\overline{M}_c$ (g/mol)	2174	2136	2186	2212	2236	2352
Measured $\overline{M}_c$ (g/mol)	1268	1815	1096	1208	1242	—
Interaction parameter, χ	0.591	0.595	0.582	0.667	0.700	0.763
Solubility parameter, δ	19.0	19.0	19.0	18.7	18.6	18.4
Volume fraction polymer, ϕ_2	0.360	0.375	0.355	0.438	0.468	0.525

[a]Data previously reported [96].

Table 5. Mechanical and physical properties of polyurethane hydrogels based on PEG 1500 and PPG 2000 [(PEG + PPG/TMP/H_{12}MDI = 1.5/1/2.5 equivalent ratio] prepared by solution polymerization in 1,4-dioxane. Solid/solvent weight ratio = 60/40.

Designation	1	2	3	4	5	6	7
Polyurethane composition (eq.)							
H_{12}MDI	2.5	2.5	2.5	2.5	2.5	2.0	2.0
PEG 1500	1.5	1.3	1.0	0.7	0.6	0.4	0.0
PPG 2000	0.0	0.2	0.5	0.8	0.9	0.6	1.0
TMP	1.0	1.0	1.0	1.0	1.0	1.0	1.0
Hydrogel properties[a]							
% H_2O	70.6	66.3	54.8	48.2	43.4	38.3	0.00
% Saline	69.2	65.6	55.2	47.1	41.0	37.8	—
% Elongation	116	100	242	487	712	231	—
Tensile strength (MPa)	0.84	0.28	0.72	1.21	1.09	3.60	—
Hardness (Shore A)	50	50	39	41	33	30	—
Appearance (c = clear, h = hazy)	h	h	c	c	c	c	c
%Non-freezing water	—	—	87.4	87.8	84.6	—	—
Physical properties of polyurethane							
Density (g/mL)	1.160	1.152	1.136	1.066	1.054	1.094	—
Moles EO	51.2	44.4	34.2	23.8	20.4	13.6	—
Moles H_2O	400	338	218	176	147	83.4	—
H_2O EO mole ratio	7.81	7.64	6.41	7.36	7.17	6.11	—
Theoretical $\overline{M}_c$ (g/mol)	2968	3068	3218	3368	3414	2386	
Interaction parameter, χ	0.507	0.557	0.668	0.749	0.804	0.846	—
Solubility parameter, δ	19.3	19.1	18.7	18.4	18.3	18.14	—
Volume fraction polymer, ϕ_2	0.263	0.305	0.420	0.501	0.552	0.594	

[a]Data previously reported [96].

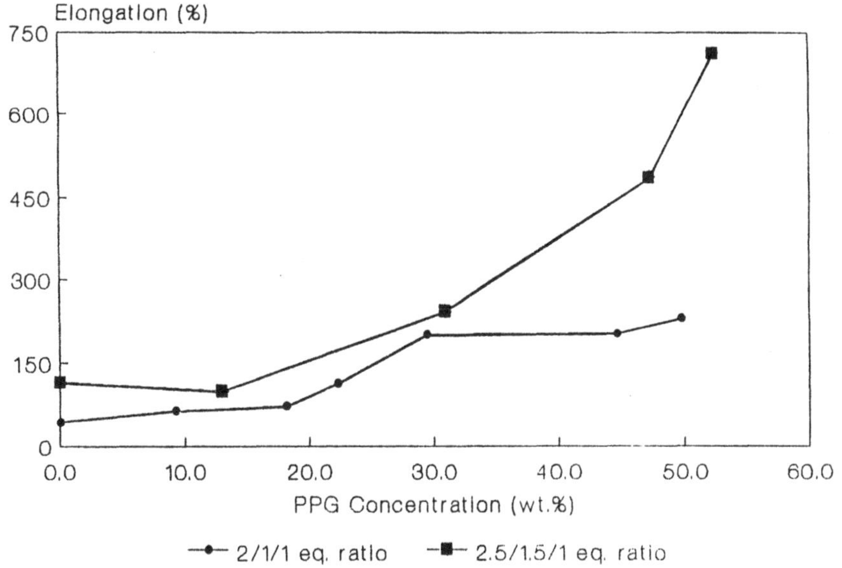

Figure 6. Effect of the PPG concentration on the elongation of polyurethane hydrogels (H_{12}MDI/(PEG 1500 + PPG 2000)/TMP equivalent ratios as indicated) prepared in solution using dioxane as solvent (40 weight percent solids).

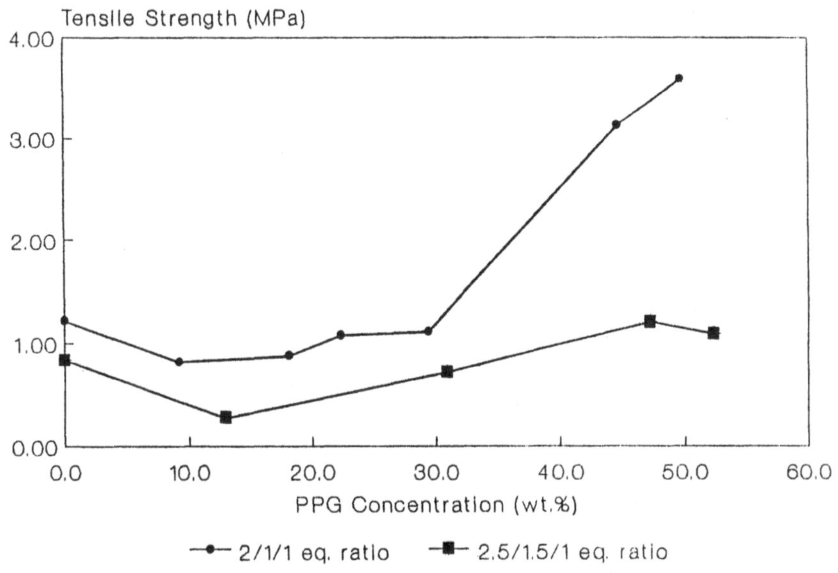

Figure 7. Effect of the PPG concentration on the tensile strength of polyurethane hydrogels (H_{12}MDI/(PEG 1500 + PPG 2000)/TMP equivalent ratios as indicated) prepared in solution using dioxane as solvent (40 weight percent solids).

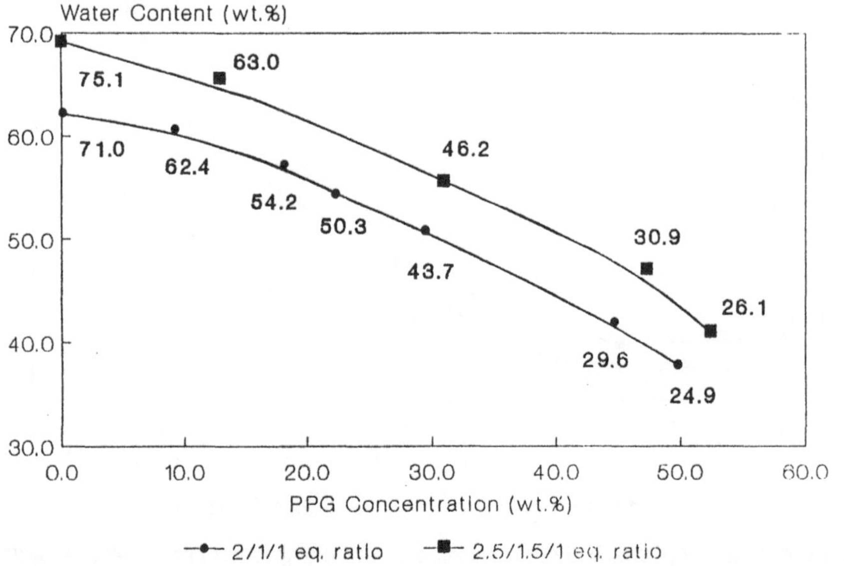

Figure 8. Effect of the PPG concentration on the water uptake of polyurethane hydrogels ($H_{12}MDI$/(PEG 1500 + PPG 2000)/TMP equivalent ratios as indicated) prepared in solution using dioxane as solvent (40 weight percent solids). Bold-faced numbers on graph indicate PEG concentration (wt%).

cally the same. The "X" in the repeat unit represents a hydrogen atom in the case of PEG and a methyl group in the case of PPG. For PEG 1500, the number of repeat units is 34.09 and for PPG 2000 this number is 34.48. This means that the overall length of the chain does not change by substituting PEG 1500 with PPG 2000. For this reason the mole ratio of water/oxyethylene groups remained relatively constant and independent of the concentration of PEG (Tables 4 and 5). The mole ratio of water/oxyethylene was between 5.2 and 7.0 for hydrogels of the more crosslinked system (Table 4), and between 6.4 and 7.8 for hydrogels of the less crosslinked system (Table 5). These results indicate that the length of the chain between crosslinks controls the water/oxyethylene mole ratio. The network which is crosslinked to a lower degree (2.5/1.5/1), showed water absorption values about 5 to 7 percent higher than hydrogels prepared at the higher degree of crosslinking (2/1/1) (Tables 4 and 5, Figure 8).

The measured $\overline{M}_c$ values were generally higher than the calculated values (Table 5). A higher experimental $\overline{M}_c$ than the theoretical value could indicate that fewer crosslinks formed than the predicted number.

The reaction of an isocyanate group with moisture present in the reaction vessel is one possible explanation for this.

The polymer-solvent interaction parameter, χ, increased from 0.591 for the hydrogel based on PEG 1500 alone, to 0.846 for the most hydrophobic hydrogel of the series with PEG/PPG ratio 2/3 (Table 4). Similarly, the χ parameter for the most hydrophilic hydrogel of the less crosslinked system was 0.507 compared to 0.804 for the most hydrophobic hydrogel of the series (Table 5). A χ parameter higher than 0.5 indicates that the polymer is crosslinked. As the χ parameter decreases, the swelling of the hydrogels increases. A χ parameter less than 0.5 indicates that the polymer is soluble [88,89].

The change in the χ parameter of the system indicates that the polarity of the polyurethane is changing. An increase in the χ parameter indicates that the system is less polar (more hydrophobic), as expected with increasing PPG concentration. The change in polarity, however, did not have a significant effect on the water/oxyethylene mole ratio. This result may be an indication that all water in the system is coordinated, bound water, since the water/oxyethylene mole ratio changed only by increasing the length of the chain between crosslinks (Tables 4 and 5).

As mentioned earlier, the water/oxyethylene mole ratio was between 6.4 and 7.8 for the less crosslinked system (1/1.5/2.5) and between 5.2 and 7.0 for the more crosslinked system (1/1/2). It was determined experimentally by DSC that a very high concentration of non-freezing (bound) water exists in these hydrogels (Tables 4 and 5). The fraction of water contained by the hydrogel which was determined to be non-freezing was 89% when the ratio of PEG/PPG was 3/1 and 96% when the PEG/PPG ratio was 2/1 in the more crosslinked system (2/1/1) (Table 5). Similarly, in the less crosslinked system (2.5/1.5/1), the fraction of water contained by the hydrogels which was determined to be non-freezing was 87% (PEG/PPG = 2/1); 88% (PEG/PPG = 7/8); and 85% (PEG/PPG = 3/2).

Although the difference in the concentration of non-freezing water between the systems of different crosslink densities is small, it would be expected that the less crosslinked system (2.5/1.5/1) which swells from 5 to 7% more, should contain less non-freezing water. With higher $\overline{M}_c$, the network should be more elastic resulting in greater free volume and more space for unbound water, as demonstrated by the experimental results.

The actual concentration of bound, non-freezing water may be even higher than these values. It is possible that surface water was present during the measurement and contributed to the free (unbound) water concentration. In addition, the heating rate of 5°C/min for the DSC measurement may have been too high to accurately observe the fraction of water which melts below 0°C.

The solubility parameters, δ's, of the polyurethane elastomers were between 18.3 and 19.3 for elastomers of both degrees of crosslinking (Tables 4 and 5). Water has an assigned solubility parameter of 23.4 and methanol 14.5 [91,97]. Polyurethane elastomers exhibited the highest swelling in formamide, which has an assigned solubility parameter of 19.2 [91,97] (Tables 6 and 7, Figure 9). The swelling of the polyurethanes in water increased as the solubility parameter of the polyurethane increased from 18.3 to 19.3, as would be expected since the solubility parameter of the polyurethane approaches the solubility parameter of water (23.4). The swelling data from polyurethane elastomers swollen in water, methanol, and formamide confirm that the solubility parameters of these polyurethane materials should be around 19, as calculated from the experimental data.

The Effect of the Hard and Soft Segments on the Swelling of Polyurethane Hydrogels

The swelling of the hydrogels is also controlled by the symmetry of the hard segments. Hydrogels prepared using IPDI exhibited higher swelling than hydrogels prepared with H_{12}MDI when the degrees of crosslinking and soft segment compositions were the same (Tables 4, 5, 8, and 9, Figure 10). The isocyanate functional groups on the H_{12}MDI

Table 6. Swelling ratios and solubility parameters of polyurethane elastomers in water, methanol and formamide.

Designation	1	2	3	4	5	6
Polyurethane composition (eq.)						
H_{12}MDI	2.0	2.0	2.0	2.0	2.0	2.0
PEG 600	1.0	—	—	—	—	—
PEG 1000	—	0.9	0.8	0.9	0.8	—
PEG 1500	—	—	—	—	—	1.0
PPG 1000	—	0.1	0.2	—	—	—
PPG 2000	—	—	—	0.1	0.2	—
TMP	1.0	1.0	1.0	1.0	1.0	1.0
Solubility parameter, δ	18.3	18.6	18.5	18.6	18.4	19.0
Swelling ratio						
In H_2O	0.48	0.95	0.83	0.96	0.77	1.59
In CH_3OH	1.02	1.28	1.33	1.59	1.50	1.50
In $HCONH_2$	1.07	1.59	1.45	1.69	1.43	2.41

Table 7. Swelling ratios and solubility parameters of polyurethane elastomers in water, methanol and formamide.

Designation	1	2	3	4	5	6
Polyurethane composition (eq.)						
$H_{12}MDI$	2.0	2.0	2.0	2.0	2.5	2.5
PEG 1500	0.9	0.8	0.9	0.8	1.5	1.3
PPG 1000	0.1	0.2	—	—	—	—
PPG 2000	—	—	0.1	0.2	—	0.2
TMP	1.0	1.0	1.0	1.0	1.0	1.0
Solubility parameter, δ	19.0	18.8	19.0	19.0	19.3	19.1
Swelling ratio						
In H_2O	1.43	1.22	1.32	1.56	2.40	1.97
In CH_3OH	1.43	1.24	1.35	2.25	2.51	2.73
In $HCONH_2$	2.13	1.79	1.96	2.43	3.95	3.16

molecule are symmetrical relative to each other and therefore yield symmetric urethane linkages when reacted with a difunctional polyol or chain extender. The position of the isocyanate groups on the IPDI molecule are asymmetric relative to each other, however, and lead to branched urethane linkages. The asymmetric branching of the urethane groups around the IPDI molecule reduces interchain interactions in the hard segments and increases the free volume of the polymer. These effects should lead to higher swelling.

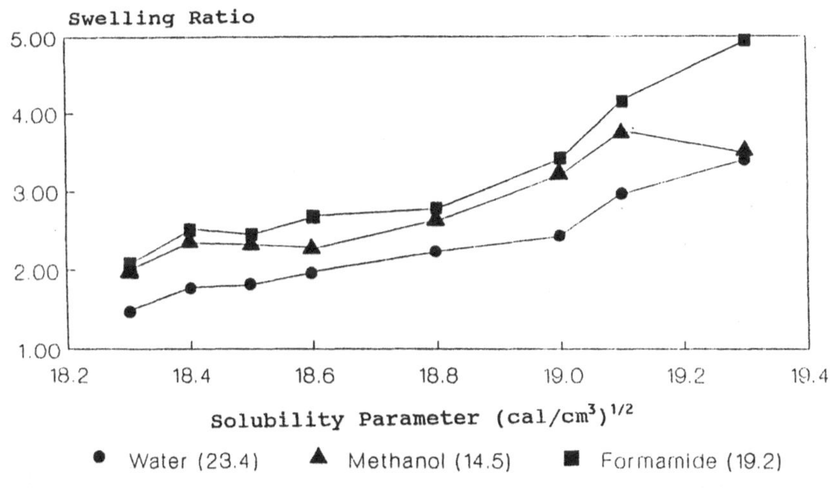

Figure 9. Effect of the solubility parameter (δ) of polyurethane elastomers ($H_{12}MDI$/PEG/TMP) on the swelling ratio of the elastomers in water, in methanol, and in formamide.

Hydrogels prepared at an equivalent ratio of IPDI/(PEG 1500 + PPG 2000)/TMP = 2/1/1 exhibited swelling similar to that obtained using $H_{12}MDI$ at an equivalent ratio of $H_{12}MDI$/(PEG 1500 + PPG 2000)/TMP = 2.5/1.5/1 (Tables 5 and 8, Figure 10). Hydrogels prepared at an equivalent ratio of 2.5/1.5/1 with IPDI exhibited higher swelling than those based on $H_{12}MDI$ (Tables 5 and 9, Figure 10).

The water content of the hydrogels decreased for both diisocyanates as the PPG concentration increased (Tables 4, 5, 8, and 9, Figure 10). With decreasing crosslink density, the water content of the hydrogels increased, as would be expected. For hydrogels based on IPDI, the mole ratio of water/oxyethylene ranged from 6.9 to 9.6 for hydrogels of the more crosslinked system (2/1/1), and from 8.1 to 10.3 for the less crosslinked hydrogels (2.5/1.5/1) (Tables 8 and 9). These results support the conclusion that the chain length between crosslinks controls the swelling of the hydrogels.

Compared to hydrogels based on $H_{12}MDI$ of similar $\overline{M}_c$, the IPDI-based hydrogels exhibited larger water/oxyethylene mole ratios (Tables

Table 8. Mechanical and physical properties of polyurethane hydrogels based on IPDI [(PEG + PPG)/TMP/IPDI = 1/1/2] prepared by solution polymerization in 1,4-dioxane. Solid/solvent weight ratio = 60/40.

Designation	1	2	3
Polyurethane composition (eq.)			
IPDI	2.0	2.00	2.0
PEG 1500	0.8	0.47	0.4
PPG 2000	0.2	0.53	0.6
TMP	1.0	1.00	1.0
Hydrogel properties[a]			
%Saline	68.9	49.6	42.1
%Elongation	96	166	282
Tensile strength (MPa)	0.39	0.48	0.75
Hardness (Shore A)	24	15	20
Appearance	clear	clear	clear
Physical properties			
Density (g/ml)	1.159	1.106	1.078
Moles EO	27.2	15.9	13.6
Moles H_2O	262	126	94.2
H_2O/EO Mole Ratio	9.63	7.89	6.91
Theoretical $\overline{M}_c$ (g/mol)	2106	2272	2306
Interaction parameter, χ	0.488	0.695	0.768
Solubility parameter, δ	19.4	18.6	18.4
Volume fraction polymer, ϕ_2	0.280	0.462	0.530

[a]Data previously reported [96].

Table 9. Mechanical and physical properties of polyurethane hydrogels based on IPDI [(PEG + PPG)/TMP/IPDI equivalent ratio = 1.5/1/2.5] prepared by solution polymerization in 1,4-dioxane. Solid/solvent weight ratio = 60/40.

Designation	1	2	3	4	5
Polyurethane composition (eq.)					
IPDI	2.5	2.5	2.5	2.5	2.5
PEG 1500	1.5	1.3	1.0	0.7	0.6
PPG 2000	—	0.2	0.5	0.8	0.9
TMP	1.0	1.0	1.0	1.0	1.0
Hydrogel properties[a]					
%H_2O	—	—	61.1	53.9	53.1
%Saline	75.4	74.3	62.3	54.0	48.9
%Elongation	weak	277	443	248	629
Tensile strength (MPa)	weak	0.30	0.79	0.78	1.34
Hardness (Shore A)	40	30	33	20	5
Appearance (h = hazy, c = clear)	h	h	c	c	c
Physical properties					
Density (g/mL)	—	—	1.135	1.121	1.058
Moles EO	—	—	34.2	23.8	20.4
Moles H_2O	—	—	274	214	210
H_2O/EO mole ratio	—	—	8.05	8.95	10.3
Theoretical $\overline{M}_c$ (g/mol)	—	—	3118	3268	3318
Interaction parameter, χ	—	—	0.611	0.681	0.803
Solubility parameter, δ	—	—	18.9	18.7	18.6
Volume fraction polymer, ϕ_2	—	—	0.359	0.432	0.454

[a]Data previously reported [96].

5 and 9). These results can be attributed to the asymmetrical, branched structure of IPDI which leads to a less ordered network. It is, therefore, not only the length of the chain between crosslinks and concentration of PEG which determine the swelling, but also the morphology of the polymer.

The polymer-water interaction parameters, χ's, for hydrogels based on IPDI were similar in magnitude to those of the hydrogels based on $H_{12}MDI$ (Tables 4, 5, 8, and 9). As the water content of the hydrogels increased, due to decreasing PPG concentration, the χ parameters decreased, as expected.

The solubility parameters for hydrogels based on IPDI were between 18.4 and 19.4, which are consistent with the values of the solubility parameters determined for $H_{12}MDI$-based hydrogels (Tables 4, 5, 8, and 9).

The number of crosslinks within the polyurethane network is controlled by the concentration of triol. For hydrogels prepared from $H_{12}MDI$, PEG 1500, PPG 2000, and TMP (PEG/PPG equivalent

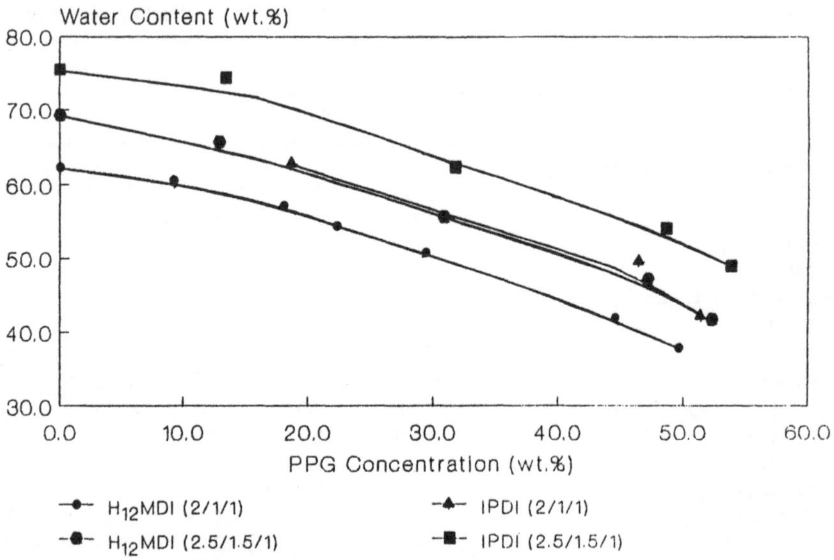

Figure 10. Effect of the diisocyanate structure and the PPG concentration on the water uptake of polyurethane hydrogels (H_{12}MDI/(IPDI)/(PEG 1500 + PPG 2000)/TMP equivalent ratios as indicated) prepared in solution using dioxane as solvent (40 weight percent solids).

ratio = 2/1), the number of crosslinked segments was increased by increasing the concentration of TMP and H_{12}MDI (Table 10, Figure 11). The water content of the hydrogels decreased from 71% to 51% by increasing the concentration of TMP from 1.5 to 4 wt%, corresponding to changing the TMP/(PEG 1500 + PPG 2000) equivalent ratio from 1/3 to 1/1. The concentration of PEG decreased by only 5% with the introduction of more crosslinker.

Although the overall concentration of PEG changed very little by increasing the concentration of TMP, the molecular weight between crosslinks ($\overline{M}_c$) increased from 1126 to 3055 g/mol (Table 10). The mole ratio of water/oxyethylene increased from 5.2 to 11.4 with increasing $\overline{M}_c$, indicating, once again, that the length of the chain between crosslinks controls the swelling.

Although the overall ratio of PEG/PPG was held constant, the hydrophobicity of the polymer network increased with increasing crosslink density (concentration of TMP), which resulted in larger values for the χ parameters. The solubility parameters for these hydrogels were between 18.6 and 19.1, similar to other systems (Tables 4, 5, 8, and 9).

As mentioned, the length of the soft segments, as well as the concentration of oxyethylene units determine the swelling of the hydrogels. For hydrogels prepared from mixtures of PEG and PPG, the equilibrium water content increased with increasing length of the soft segment, as the molecular weight of PEG increased from 1000 to 1500 and the molecular weight of PPG increased from 1000 to 2000 (Tables 11 and 12, Figure 12). The highest degree of swelling was obtained using the combination of PEG 1500 and PPG 2000, which corresponds to the highest $\overline{M}_c$ value, although the oxyethylene group concentration for some of the hydrogels of this series was lower than for hydrogels with lower $\overline{M}_c$ which swelled less.

The lowest swelling and water/oxyethylene mole ratio was obtained for the hydrogel based on PEG 600, which correlates to the lowest $\overline{M}_c$ value (593 g/mol) (Table 11). The water/oxyethylene mole ratio of 2.3 possibly indicates that all water contained by the hydrogel is bound

Table 10. Mechanical and physical properties of polyurethane hydrogels based on PEG 1500 and PPG 2000 (PEG/PPG = 2/1 equivalent ratio) prepared by solution polymerization in 1,4-dioxane. Solid/solvent weight ratio = 60/40.

Designation	1	2	3	4	5
Polyurethane composition (eq.)					
%H_{12}MDI	2.00	2.25	2.50	2.88	4.00
PEG 1500	0.67	0.83	1.00	1.25	2.00
PPG 2000	0.33	0.42	0.50	0.63	1.00
TMP	1.00	1.00	1.00	1.00	1.00
Hydrogel properties[a]					
H_2O	50.2	56.8	54.8	61.8	69.6
%Saline	50.8	57.0	55.6	61.2	71.6
%Elongation	201	467	242	997	99
Tensile strength (MPa)	1.12	1.52	0.72	1.19	3.53
Hardness (Shore A)	38	33	39	17	9
Appearance (c = clear)	c	c	c	c	c
Physical properties of polyurethane					
Density (g/mL)	1.144	1.125	1.119	1.237	1.159
Moles EO	22.8	28.4	34.2	42.6	68.2
Moles H_2O	118	202	218	356	780
H_2O/EO Mole Ratio	5.15	7.10	6.42	8.35	11.4
Theoretical $\overline{M}_c$ (g/mol)	2484	2736	3218	3940	6110
Measured $\overline{M}_c$ (g/mol)	1242	1360	1526	—	—
Interaction parameter, χ	0.700	0.652	0.672	0.595	0.568
Solubility parameter, δ	16.6	18.8	18.7	19.0	19.1
Volume fraction polymer, ϕ_2	0.464	0.409	0.423	0.333	0.273

[a]Data previously reported [96].

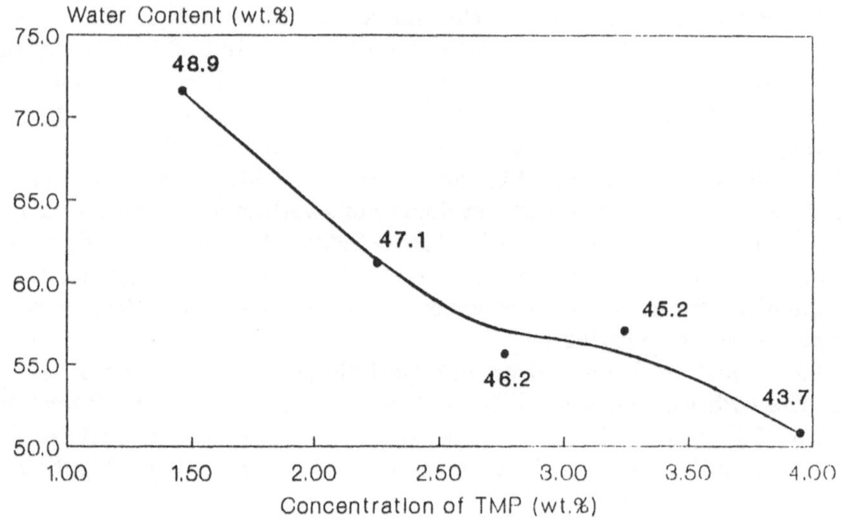

Figure 11. Effect of the crosslink concentration on the water uptake of polyurethane hydrogels [H_{12}MDI/(PEG 1500 + PPG 2000)/TMP] prepared in solution using dioxane as solvent (40 weight percent solids). Bold-faced numbers on graph indicate PEG concentration (wt%).

water. Water binding to PEG results in a very organized, crystalline structure [6,87] which may explain the haziness of this hydrogel. By increasing the molecular weight of the PEG to 1000 and introducing PPG 1000 into the system, the water/oxyethylene mole ratio increased to 4. Hydrogels with water coordination numbers of 4 were also hazy, with one exception; the hydrogel containing the highest concentration of PPG 2000 (PEG/PPG = 4/1) was clear. Increasing the molecular weight of the PEG to 1500 resulted in a water/oxyethylene mole ratio of 5. All hydrogels with water coordination numbers of 5 were clear. This result suggests that there may be a higher concentration of unbound water present in these hydrogels. However, there exists the possibility that all water is bound in a well-organized phase, since the PPG by itself might disrupt the crystallinity of the network.

The water-polymer interaction parameters again reflect the polarity of the soft segments by increasing with increasing PPG concentration (Tables 11 and 12). It is interesting to note that the χ parameter for the hydrogel based on PEG 600 (100% oxyethylene-based soft segments) was higher than any of the χ parameters for hydrogels containing PPG. This result indicates that this polyurethane material is less hydro-

Table 11. Mechanical and physical properties of polyurethane hydrogels based on a mixture of PEG and PPG polyols prepared by solution polymerization in 1,4-dioxane. Solid/solvent weight ratio = 60/40.

Designation	1	2	3	4	5
Polyurethane composition (eq.)					
$H_{12}MDI$	2.0	2.0	2.0	2.0	2.0
PEG 600	1.0	—	—	—	—
PEG 1000	—	0.9	0.8	0.9	0.8
PPG 1000	—	0.1	0.2	—	—
PPG 2000	—	—	—	0.1	0.2
TMP	1.0	1.0	1.0	1.0	1.0
Hydrogel properties[a]					
%H_2O	32.6	48.7	45.5	49.1	43.6
%Saline	19.5	51.6	49.3	50.9	47.6
%Elongation	303	33	43	77	60
Tensile strength (MPa)	2.43	0.81	0.86	1.19	0.88
Hardness (Shore A)	—	45	50	50	50
Appearance (h = hazy, c = clear)	h	h	h	h	c
Physical properties of polyurethane					
Density (g/mL)	1.441	1.152	1.149	1.107	1.113
Moles EO	13.6	20.4	18.2	20.4	18.2
Moles H_2O	32.6	85.0	74.8	91.8	77.8
H_2O/EO mole ratio	2.39	4.16	4.11	4.49	4.28
Theoretical $\bar{M}_c$ (g/mol)	1186	1586	1586	1686	1786
Interaction parameter, χ	0.795	0.690	0.727	0.702	0.766
Solubility parameter, δ	17.9	—	18.6	18.5	18.4
Volume fraction polymer, ϕ_2	0.588	0.477	0.510	0.483	0.537

[a]Data previously reported [96].

Table 12. ***Mechanical and physical properties of polyurethane hydrogels based on a mixture of PEG and PPG polyols prepared by solution polymerization in 1,4-dioxane. Solid/solvent weight ratio = 60/40.***

Designation	**1**	**2**	**3**	**4**
Polyurethane composition (eq.)				
$H_{12}MDI$	2.0	2.0	2.0	2.0
PEG 1500	1.0	0.9	0.8	0.9
PPG 1000	—	0.1	0.2	—
PPG 2000	—	—	—	0.1
TMP	1.0	1.0	1.0	1.0
Hydrogel properties[a]				
$\%H_2O$	61.5	58.8	54.9	56.8
%Saline	62.2	58.0	56.6	60.6
%Elongation	164	43	51	65
Tensile strength (MPa)	1.22	1.08	1.08	0.82
Hardness (Shore A)	60	50	55	55
Appearance (c = clear)	c	c	c	c
Physical properties of polyurethane				
Density (g/ml)	1.109	1.137	1.177	1.261
Moles EO	34.0	30.6	27.2	30.6
Moles H_2O	187	164	136	158
H_2O/EO mole ratio	5.49	5.34	4.98	5.16
Theoretical $\overline{M}_c$ (g/mol)	2086	2036	1986	2136
Measured $\overline{M}_c$ (g/mol)	1087	1564	2055	1815
Interaction parameter, χ	0.591	0.604	0.632	0.595
Solubility parameter, δ	19.0	19.0	18.8	19.0
Volume fraction polymer, ϕ_2	0.360	0.380	0.411	0.375

[a]Data previously reported [96].

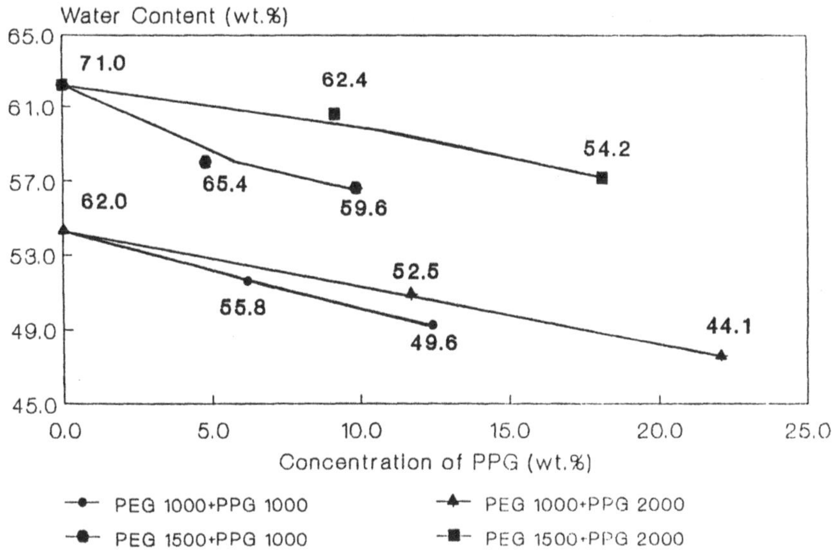

Figure 12. Effect of the soft segment molecular weight and PPG concentration on the water uptake of polyurethane hydrogels ($H_{12}MDI/(PEG + PPG)/TMP = 2/1/1$ equivalent ratio) prepared in solution using dioxane as solvent (40 weight percent solids). Bold-faced numbers on graph indicate PEG concentration (wt%).

philic due to the high concentration of TMP crosslinks which results in a lower $\overline{M}_c$.

The solubility parameters of these hydrogels ranged from 18.4 to 19.0, with the exception of the hydrogel based on PEG 600 which had a solubility parameter of 17.9. The lower δ value indicates a less polar, more hydrophobic polymer.

Comparison of hydrogels prepared from $H_{12}MDI$, PEG, and TMP with hydrogels prepared from IPDI, PEG, and P-726, shows that the same degree of swelling was possible at 50% lower PEG concentration by using a higher molecular weight crosslinker, which increases the $\overline{M}_c$ of the network (Tables 2 and 13, Figure 13). This difference in swelling between the $H_{12}MDI$ and IPDI-based elastomers cannot be attributed to the difference in isocyanate structure, since it was previously shown that hydrogels based on these two isocyanates, the same polyol composition, and crosslinker exhibited only about a 5% difference in swelling (Tables 4, 5, 8, and 9, Figure 10).

The difference in swelling is a result of the increase in chain length between crosslinks, as was demonstrated in previous systems. By substituting the higher molecular weight triol (P-726) for TMP as a cross-

Table 13. Formulations and properties of polyurethane hydrogels based on PEG, IPDI, and P-726 prepared by solution polymerization in 1,4-dioxane. Solid/solvent weight ratio = 60/40.

Designation	1	2	3
Polyurethane composition (eq.)			
IPDI	2.0	2.0	2.0
PEG 600	1.0	—	—
PEG 1000	—	1.0	—
PEG 1500	—	—	1.0
P-726	1.0	1.0	1.0
Hydrogel properties[a]			
%Saline	25.9	50.9	67.1
%Elongation	469	466	1172
Tensile strength (MPa)	1.05	1.14	0.28
Hardness (Shore A)	8	7	3
Appearance (c = clear)	c	c	c
Physical properties of polyurethane			
Moles EO	13.6	22.6	34.0
Moles H_2O	57.8	195	440
H_2O/EO mole ratio	4.24	8.61	12.9
Theoretical $\overline{M}_c$ (g/mol)	2978	3378	3878

[a]Data previously reported [96].

linker, the $\overline{M}_c$ of the polymer network increased from 593 g/mol for PEG 600/TMP, to 1489 g/mol for PEG 600/P-726; from 793 g/mol for PEG 1000/TMP, to 1689 g/mol for PEG 1000/P-726; and from 1043 g/mol for PEG 1500/TMP to 1939 g/mol for PEG 1500/P-726 (Tables 12 and 13).

The elongation of the hydrogels crosslinked with P-726 increased with increasing $\overline{M}_c$ from 500 to 1200%, while the water content of the gels increased from 25 to 70%. The elongation values of the hydrogels prepared with TMP ranged from 35 to 100%. The high elongation of the hydrogels at high water content could offer specific advantages to applications such as contact lenses where a high concentration of water in the gel is a requirement while maintaining structural integrity of the material.

The swelling of hydrogels crosslinked to a lower degree with P-726, which exhibited similar swelling to hydrogels prepared with TMP at much higher concentrations of oxyethylene units, indicates that there must be both bound and unbound water present in the hydrogel. For hydrogels crosslinked with TMP, the mole ratio of water/oxyethylene increased from 1.60 for PEG 600, to 3.26 for PEG 1000, and to 6.79 for

PEG 1500. The hydrogels crosslinked with P-726 exhibited higher water/oxyethylene mole ratios: 4.24 for PEG 600; 8.61 for PEG 1000; and 12.9 for PEG 1500. These results suggest that hydrogels based on P-726 must contain a significantly higher fraction of unbound (unfreezing) water than hydrogels based on TMP.

The effect of the elasticity of the chain between covalently bonded crosslinks on the swelling of the hydrogels was investigated using ethylene glycol as a chain extender and P-726 as crosslinker (Table 14, Figures 14 and 15). The hard segment concentration refers to the concentration of diisocyanate and chain extender (EG) for the system IPDI/PEG 1500/P-726/EG with the equivalent ratio changing from 2/1/1/0 to 12/1/1/10. The water content of the hydrogels decreased from 67 to 55% by doubling the concentration of hard segments from 11 to 22%. Further increases in the hard segment concentration up to 50 wt% had little effect on the swelling.

The addition of ethylene glycol into the system has the effect of increasing the chain length of the segment between covalently bonded crosslinks. The $\bar{M}_c$ of the polymer network increased from 1939 g/mol

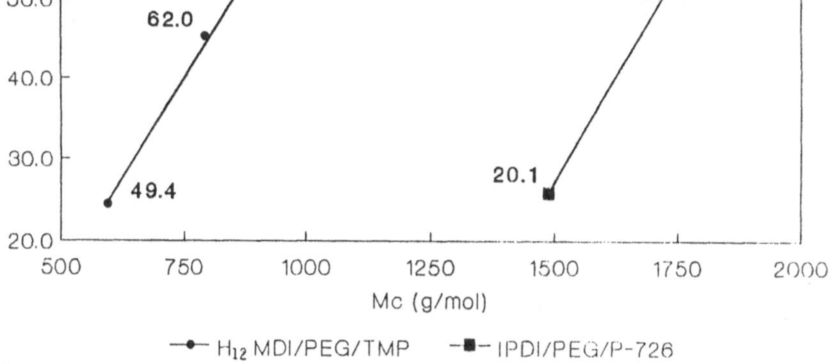

Figure 13. Effect of the molecular weight between crosslinks on the water uptake of polyurethane hydrogels ($H_{12}MDI$/PEG/TMP and IPDI/PEG/P-726) prepared in solution using ethoxyethyl-acetate as solvent (40 weight percent solids). Bold-faced numbers on graph indicate PEG concentration (wt%).

Table 14. Formulations and properties of polyurethane hydrogels based on IPDI, PEG 1500, P-726, and ethylene glycol prepared by solution polymerization in 1,4-dioxane. Solid/solvent weight ratio = 60/40.

Designation	1	2	3	4	5	6
Polyurethane composition (eq.)						
IPDI	2.0	3.0	4.0	5.0	7.0	12.0
PEG 1500	1.0	1.0	1.0	1.0	1.0	1.0
P-726	1.0	1.0	1.0	1.0	1.0	1.0
EG	—	1.0	2.0	3.0	5.0	10.0
Hydrogel properties[a]						
%Saline	67.1	69.9	54.6	56.6	52.5	58.3
%Elongation	1172	1235	522	1127	770	519
Tensile strength (MPa)	0.28	1.48	1.04	0.69	1.61	5.17
Hardness (Shore A)	3	0	15	8	19	60
Appearance (c = clear, h = hazy)	c	c	c	c	c	c
Physical properties of polyurethane						
Moles EO	34.0	34.0	34.0	34.0	34.0	34.0
Moles H_2O	440	536	298	342	326	522
H_2O/EO mole ratio	14.9	15.8	8.76	10.1	9.59	15.3
Theoretical $\overline{M}_c$ (g/mol)	3878	4162	4446	4730	5298	6718

[a]Data previously reported [96].

for the hydrogel containing no ethylene glycol, to 3359 g/mol for the hydrogel containing 10 equivalents of ethylene glycol. The fact that the water concentration of the hydrogels remained relatively constant between 2 equivalents and 10 equivalents of ethylene glycol indicates that swelling of hydrogen-bonded hard segments does not occur.

By adding ethylene glycol to the system as a chain extender, the $\overline{M}_c$ of the polymer network increased more than in previous systems when the $\overline{M}_c$ was controlled by the crosslink density. Although the overall length of the chain between crosslinks increased, the elasticity of the chain was reduced. It is, therefore, not only the length of the chain between crosslinks which determines the swelling, but also the chain elasticity (Figures 14 and 15).

The tensile strength increased from 0.3 MPa to 5 MPa, when the hard segment concentration approached 50 wt%, confirming that swelling of the hydrogen-bonded hard segments did not occur (Table 14). If swelling of these segments had occurred, there would be no interaction between adjacent chains and no contribution to the tensile strength.

The elongation of the hydrogels decreased from 1200% to 500% with the addition of 10 equivalents of ethylene glycol to the system (Table 14). The decrease in elongation also indicates that the overall elasticity

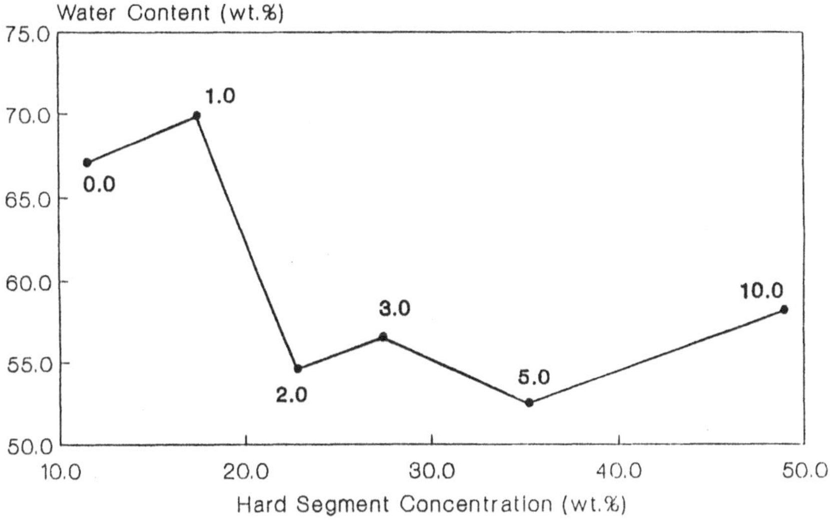

Figure 14. *Effect of the hard segment concentration on the water uptake of polyurethane hydrogels (IPDI/PEG 1500/P-726 = 2/1/1 equivalent ratio) prepared in solution using dioxane as solvent and ethylene glycol as chain extender (40 weight percent solids). Bold-faced numbers on graph indicate equivalents of ethylene glycol per equivalent of P-726 triol.*

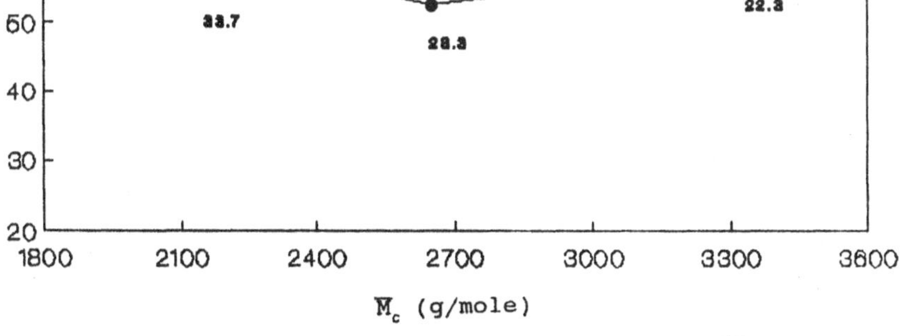

Figure 15. *Effect of the molecular weight between crosslinks ($\overline{M}_c$) on the water uptake of polyurethane hydrogels (IPDI/PEG/P-726/EG) prepared in solution using dioxane as solvent (40 weight percent solids). Bold-faced numbers on graph indicate concentration of PEG (wt%).*

of the polymer network decreases with the addition of hydrogen-bonded hard segments.

CONCLUSIONS

The concentration of hydrophilic groups in the polyurethane network by itself does not determine the swelling of the hydrogels. The polymer morphology contributes significantly to the swelling. Increasing the length of the polymer chain between crosslinks, $\bar{M}_c$, while maintaining the same concentration of oxyethylene groups, increased the water absorption significantly. However, if the increase in $\bar{M}_c$ was achieved through the introduction of rigid segments, the swelling did not increase, reflecting the importance of chain elasticity on swelling.

The solubility and polymer-solvent interaction parameters reflect the compatibility of the polyurethane with water and can be utilized to select hydrogels for specific applications. However, there exists a limited number of reference values for the solubility parameters of hydrophilic polyurethanes in the literature. The solubility parameters for the hydrophilic polyurethanes studied were in the range of 19.0 to 19.3 $(cal/cm^3)^{1/2}$. The measured solubility parameters were confirmed by swelling of the polyurethanes in solvents of known solubility parameters.

REFERENCES

1. Kiraly, R. J. and Y. Nose. 1974. *Biomater. Med. Devices Artif. Organs.*, 2:207.
2. Kudela, V. 1986. In *Encyclopaedia of Polymer Science and Engineering*, 2nd ed., New York: John Wiley and Sons, Vol. 7, pp. 783–807.
3. Hoffman, A. S. 1982. In *Macromolecules*, H. Benoit and R. Rempp, eds., New York: Pergamon, pp. 321–335.
4. Kaganov, A. L., J. Stamberg and P. Synek. 1976, *J. Biomed. Mater. Res.*, 10:1.
5. Tanaka, T. 1986. In *Encyclopaedia of Polymer Science and Engineering*, 2nd ed., New York: John Wiley and Sons, Vol. 7, pp. 514–531.
6. Graham, N. B. and M. E. McNeill. 1988. *Makromol. Chem., Macromol. Symp.*, 19:255–273.
7. Gould, F. E. and C. W. Johnston. 1988. U.S. Patent 4,780,512.
8. Graham, N. B. and M. E. McNeill. 1984. *Biomaterials*, 3:27–36.
9. McNeill, M. E. and N. B. Graham. 1987. In *ACS Symposium Series, Controlled Release Technology: Pharmaceutical Applications*, P. I. Lee and W. R. Good, eds., Ch. 12.

10. McNeill, M. E. and N. B. Graham. 1984. *J. Contr. Release,* 1:99–117.
11. Graham, N. B. and M. E. McNeill. 1985. *J. Contr. Release,* 2:231–244.
12. Wichterle, O. and D. Lim. 1960. *Nature,* 185:117–118.
13. Macret, M and G. Hild. 1982. *Polymer,* 23:748–753.
14. Singer, H. K. and A. R. LeBoeuf. 1986. In *Encyclopaedia of Polymer Science and Engineering,* 2nd ed., New York: John Wiley and Sons, Vol. 4, pp. 164–173.
15. Lai, Y. C. and L. J. Baccei. 1991. *J. Appl. Polym. Sci.,* 42:3173–3179.
16. Harvey, T. B., W. B. Meyers and L. M. Bowman. 1990. In *Progress in Biomedical Polymers,* C. G. Gebelein and R. L. Dunn, eds., New York: Plenum, pp. 1–5.
17. Mueller, K. F. 1991. U.S. Patent 4,983,702.
18. Arkels, B. C. 1983. *Chemtech,* 13(9):513.
19. Bowman, L. M., T. B. Harvey and W. E. Meyers. 1990. In *Progress in Biomedical Polymers,* C. G. Gebelein and R. L. Dunn, eds., New York: Plenum, pp. 7–12.
20. Meyers, W. E., T. B. Harvey and L. M. Bowman. 1990. In *Progress in Biomedical Polymers,* C. G. Gebelein and R. L. Dunn, eds., New York: Plenum, pp. 13–17.
21. Hart, D. E., R. R. Tisdale and R. A. Sack. 1986. *Ophthalmology,* 93:495.
22. White, M. L. and G. H. Davidson. 1966. *J. Polym. Sci.,* 55:731.
23. David, B. K. 1974. *Proc. Natn. Acad. Sci. USA,* 71:3120.
24. Good, W. R. 1978. In *Polymeric Delivery System,* R. J. Kostelink, ed., London: Gordon and Breach, p. 139.
25. Lee, B. H., S. H. Shim and D. J. Angrade. 1972. *Polym. Prepr.,* 13:723.
26. Hoffman, A. S., W. G. Kroft and C. Harris. 1972. *Polym. Prepr.,* 13:723.
27. Mueller, K. F. 1981. European Patent Appl. 81810300.4.
28. Davis, T. P., M. B. Huglin and C. F. Daniel. 1988. *Polymer,* 29:701–706.
29. Huglin, M. B. and M. B. Zakarta. 1983. *J. Appl. Polym. Sci.,* 28:2451.
30. Huglin, M. B. and M. B. Zakarta. 1984. *Polymer,* 25:797.
31. Huglin, M. B. and M. B. Zakarta. 1986. *J. Appl. Polym. Sci.,* 31:457.
32. Huglin, M. B., M. B. Zakarta and M. M. Rehab. 1986. *Macromolecules,* 19:1986.
33. Molyneaux, P. 1983. *Water Soluble Synthetic Polymers: Properties and Behavior,* Boca Raton, FL: CRC Press, Vol. 1, Ch. 4.
34. Blair, E. A. and D. E. Hudgin. 1974. U.S. Patent 3,786,035.
35. Gould, F. E. and C. W. Johnston. 1982. U.S. Patent 4,359,558.
36. Gould, F. E. and C. W. Johnson. 1983. U.S. Patent 4,408,023.
37. Gould, F. E. and C. W. Johnston. 1984. U.S. Patent 4,454,309. 1985; 4,496,535.
38. Gould, F. E. and C. W. Johnston. 1985. British Patent 2,150,938.
39. Gould, F. E. and C. W. Johnston. 1985. German Patent 3,344,001.
40. Gould, F. E. and C. W. Johnston. 1984. U.S. Patent 4,424,305.

41. Saunders, J. H. and K. C. Frisch. 1962–1964. *Polyurethanes Chemistry and Technology*, Parts I and II. New York: John Wiley and Sons.
42. Coury, A. J., K. E. Cobian, P. T. Calahan and A. H. Jevne. 1984. In *Advances in Urethane Science and Technology*, K. C. Frisch and D. Klempner, eds., Lancaster, PA: Technomic Publishing Co. Inc., Vol. 9, pp. 130–168.
43. Boretos, J. W. 1980. *Pure Appl. Chem.*, 52:1851.
44. Hanson, S. R., L. A. Harker, B. D. Ratner and A. S. Hoffman. 1980. *J. Lab. Clin. Med.*, 95:289.
45. Lelah, M. D., L. K. Lambrecht, B. R. Young and S. L. Cooper. 1983. *J. Biomed. Mater. Res.*, 17:1.
46. Boretos, J. W., W. S. Pierce, R. E. Baier, A. F. LeRoy and H. J. Donachy. 1975. *J. Biomed. Mater. Res.*, 9:327.
47. Boretos, J. W., D. E. Detmer and J. H. Donachy. 1971. *J. Biomed. Mater. Res.*, 5:373.
48. Kolff, J., G. Burkett and J. Feyen. 1974. *J. Biomater. Med. Devices Artif. Organs*, 1:669.
49. Nathan, A., D. Bolikal, N. Vyavahare, S. Zalipsky and J. Kohn. 1992. *J. Macromolecules*, 25:4476–4484.
50. Stone, F. W. and J. J. Stratta. 1967. In *Encyclopaedia of Polymer Science and Technology*, New York: John Wiley and Sons, Vol. 6, pp. 103–145.
51. Bailey, F. E., Jr. and J. V. Koleske. 1976. *Poly(ethylene oxide)*. New York: Academic.
52. Malcolm, G. N. and J. S. Rowlinson. 1957. *Trans. Faraday Soc.*, 53:921.
53. Hagar, S. L. and T. B. Macrury. 1980. *J. Appl. Polym. Sci.*, 25:1559–1571.
54. Bailey, F. E., Jr. and J. V. Koleske. 1967. In *Nonionic Surfactants*, M. J. Schick, ed., New York: Academic.
55. Courval, G. J. and D. G. Gray. 1983. *Polymer*, 24:323–326.
56. Graham, N. B., M. Zulfiqar, N. E. Nwachuku and A. Rashid. 1989. *Polymer*, 30:528–533.
57. Bailey, F. E., Jr. and R. W. Callard. 1959. *J. Appl. Polym. Sci.*, 1:373–374.
58. Bailey, F. E., Jr. and R. W. Callard. 1959. *J. Appl. Polym. Sci.*, 1:56.
59. Nakayama, H. 1970. *Bull. Chem. Soc., Japan*, 43:1683–1686.
60. Liu, K. J. 1968. *Macromolecules*, 1(3):213–217.
61. Liu, K. J. and J. L. Parsons. 1969. *Macromolecules*, 2(5):529–533.
62. Yoshihara, T., H. Tadakoro and S. Murahashi. 1964. *J. Chem. Phys.*, 41(9):2902–2911.
63. Rabolt, J. F., K. W. Johnson and R. N. Zitter. 1974. *J. Chem. Phys.*, 81(2):504–506.
64. Boehmke, G. and R. Heusch. 1960. *Fette, Seifen*, 92(2):87–91.
65. Andersson, B. and G. Oloffson. 1987. *Colloid & Polymer Science*, 265:318–328.
66. Maxfield, J. and I. W. Shepherd. 1975. *Polymer*, 16:505–509.
67. Koenig, J. L. and A. C. Angood. 1970. *J. Polym. Sci.*, 8:1787–1796.
68. Maconnachie, A., P. Vasudevan and G. Allen. 1978. *Polymer*, 19:33.

69. Kjellander, R. and E. Florin. 1981. *J.C.S., Trans. Faraday Soc.,* 1981. 77:2053.
70. Bogdanov, B. and M. Mihailov. 1986. *J. Macromol. Sci. Phys.,* B25:89.
71. Bogdanov, B. and M. Mihailov. 1987. *J. Macromol. Sci. Phys.,* B26:59.
72. Maron, S. H. and F. E. Filisko. 1972. *J. Macromol. Sci.,* B6:79.
73. Makayama, H. 1970. *Bull. Chem. Soc., Japan,* 43:1683.
74. Cunningham, R. G. and G. N. Malcolm. 1961. *J. Phys. Chem.,* 65:1454.
75. Graham, N. B., N. E. Nwachuku and D. J. Walsh. 1982. *Polymer,* 23:1345–1349.
76. Aladesulu, I., N. B. Graham and R. W. Richards. 1983. *Polymer,* 24:279–284.
77. Graham, N. B. and M. Zulfiqar. 1989. *Polymer,* 30:2130–2135.
78. Selegny, E., G. Boyd and H. Gregor, eds. 1976. *Charged Gels and Membranes.* Part I and II. Dordrecht, Holland: D. Reidel.
79. Castellan, G. 1983. *Physical Chemistry,* 3rd ed. Reading: Addison-Wesley, pp. 288–289.
80. Brandup, J. and E. H. Immergut. 1975. *Polymer Handbook,* 2nd ed., New York, Wiley-Interscience, Vol. 7, pp. 519–557.
81. Treloar, L. R. G. 1975. *The Physics of Rubber Elasticity,* 3rd ed., Oxford: Clarendon Press, pp. 129–159.
82. Clark, E. S. 1986. In *Encyclopaedia of Polymer Science and Engineering,* 2nd ed., New York: John Wiley and Sons, Vol. 5, pp. 365–414.
83. Tanaka, T. 1986. In *Encyclopaedia of Polymer Science and Engineering,* 2nd ed., New York: John Wiley and Sons, Vol. 7, pp. 514–531.
84. Brady, J. E. and G. E. Humiston. 1982. *General Chemistry Principles and Structure,* 3rd ed., New York: John Wiley and Sons, pp. 316–354.
85. Kwei, T. K. 1979. In *Macromolecules an Introduction to Polymer Science,* F. A. Bovey and F. H. Winslow, eds., Orlando, FL: Academic Press, pp. 273–316.
86. Treloar, L. R. G. 1975. *The Physics of Rubber Elasticity,* 3rd ed., Oxford: Clarendon Press, pp. 134–135.
87. Katz, J. R. 1925. *Chem.-Ztg.,* 49:353.
88. Munk, P. 1989. *Introduction to Macromolecular Science,* New York: John Wiley and Sons, pp. 217–230.
89. Flory, P. J. 1953. *Principles of Polymer Chemistry,* Ithaca, NY: Cornell University Press.
90. Rodriguez, F. 1982. *Principles of Polymer Systems,* 2nd ed., New York: Hemisphere, pp. 23–32.
91. Barton, A. F. M., ed. 1983. *CRC Handbook of Solubility Parameters and Other Cohesion Parameters,* Boca Raton, FL: CRC Press, pp. 85–87.
92. Huglin, M. B. and J. M. Rego. 1991. *Macromolecules,* 24:2556–2563.
93. Damusis, A., ed. 1967. *Sealants.* New York: Reinhold, pp. 25–33.
94. Cluff, E. F., E. K. Gladding and R. Pariser. 1960. *J. Polym. Sci.,* 45:341–345.
95. Cooper, A. R. 1990. *Polym. Eng. and Sci.,* 30(24):1591–1594.

96. Haschke, E., V. Sendijarevic, S. Wong, K. C. Frisch and G. Hill. 1994. *J. Elastomers and Plastics,* 26(1):41–57.
97. *CRC Handbook of Chemistry and Physics,* 1982.
98. Hepburn, C. 1982. *Polyurethane Elastomers,* New York: Applied Science, pp. 69–70.

Recycling of Polyurethane Polymers

MICHELE MODESTI
University of Padova
Industrial Chemistry–Eng. Dept.
Via Marzolo 9, I–35100 Padova, Italy

ABSTRACT

Thermoset polyurethane polymers, such as rigid and flexible foams, elastomers and RIM-PUR can be recycled. A lot of processes have been developed, based on physical and chemical methods, which produce materials that can be directly or indirectly reused in the production of new polyurethanes.

Physical recycling methods (primary and secondary recycling), including recycling of material as filler, flake and particle rebonding and hot compression molding, have found different applications, especially in the field of flexible foams and RIM-PUR.

Chemical recycling methods (tertiary recycling), such as hydrogenation, hydrolysis, glycolysis and pyrolysis seem to be more interesting, as they are able to assure a high recovery of these materials. In particular the glycolysis process has been dealt with in depth as it seems to be the most promising in versatility and cheapness among the different chemical processes.

Incineration with thermal recovery (quaternary recycling) is a possible last stage operation for those materials that cannot be recovered in other ways.

These recycling technologies contribute to overall energy savings and eliminate the cost and ecological problems of landfill.

INTRODUCTION

POLYURETHANE MATERIALS ARE COMPRISED OF POLYMERS DERIVED from the reaction between isocyanate and organic active hydrogen compounds. The high versatility of these systems, both linear and crosslinked, allows realization in a very diverse spectrum of products. Polyurethane polymers constitute the largest of the thermoset resins due to their intrinsic properties and versatility: about 3,500 million pounds of polyurethane materials were sold in the United States in

1993 [1]. The fields of largest application are: rigid foams for thermal insulation (984 million pounds), flexible foams for bedding and furniture applications (about 1.7 billion pounds), elastomer applications (about 800 million pounds, of which Reaction Injection Molding (RIM) represents 210 million pounds). Polyurethanes are also used in many other industrial applications as wood substitutes, varnish and adhesives, etc.

The ecological disposal of such wastes and scrap became a difficult problem to solve and it is presently of considerable concern.

Unlike other polymer industries, the production of the greatest part of PUR involves chemical reactions during the molding, which means a greater risk in the process. In spite of the improvement of the equipment and systems, production scrap and waste cannot be completely avoided. Besides these wastes, consideration must be given to disposal of products at the end of their lives.

Up to now, landfilling has been the principal disposal method and is not free from the danger of pollution anyway, because of possible degradation of the polymers. This method also leads to a total loss of materials.

Solutions to the problem of disposal have been the object of numerous studies, made especially urgent by the less and less availability of landfills and public opinion, which puts pressure on finding new ecological ways.

The solution to these problems can be found in the recycling of the materials which can be classified in:

- *recycling of materials:* refers here to the direct reuse of a material or its splitting-up into its basic constituents and their recycling into the production process
- *recycling of energy:* understood to mean the utilization of the high energy content of materials by conversion into heat or electricity, e.g. by combustion

These kinds of recycling may be classified, according to ASTM D5033 [2], into primary, secondary, tertiary and quaternary:

- *primary recycling:* this encompasses the recycling of waste materials in the original application. Both thermoplastic materials and some thermoset materials may be involved in this recycling. The main technical problems encountered in primary recycling are the contamination of the reprocessed plastic and the degradation of the material due to repeated processing, resulting in deterioration of properties such as appearance, chemical resistance, processability and mechanical characteristics [3].
- *secondary recycling:* this encompasses the recycling of wastes into less demanding applications, e.g., non-appearance and non-structural

parts. Secondary recycling of polyurethanes consists of different size reduction processes and is different from primary recycling in the final use of the granulated material (e.g., hot compression molding, particle rebonding).

- *tertiary recycling:* this involves the production of basic chemical monomers, low–medium molecular weight oligomers and fuels from plastic waste through thermal or chemical degradation of polymers by the action of water, organic compounds, ammonia or via biological and photolytic degradation.
- *quaternary recycling:* this involves incineration with energy recovery. It is generally applied to end-life materials and mixed plastic wastes. The combustion of polymeric materials is a complex phenomenon which involves pyrolysis processes, oxidative degradation and real combustion. Due to its complexity, the combustion must be carried out in suitable plants equipped with scrubbers in order to avoid the release of toxic gases.

RECYCLING OF MATERIALS

Physical Processes

A number of very promising physical methods for PUR recycling exists, some of which have been developed into the pilot stage, while a few are already being used in actual practice.

In primary and secondary recycling, polyurethane waste is mechanically turned into flakes, granules, or powder which are then used to produce new materials. In these processes the polymeric structure of the polyurethane waste remains substantially unchanged [4,5].

Size reduction of scrap polymers was not an economical process because it required the use of cryogenic grinding, which added a prohibitive cost. A more recent technique to economically grind the thermoset scraps into a size useful as a filler in other applications was discovered [6,7]. It uses impact disc mills [6,7] which employ air during the pulverization to keep the polymer cool and to avoid agglomeration and polymer degradation. At this step, the scrap powder is normally separated into two classifications: less and greater than about 250 μm. These classified materials can be used in different recycling routes.

Recycling of Material as Filler

One possibility for direct utilization of scrap is to use them in granular or powder form as filler in new parts. Development work has incorporated regrind into a variety of other applications such as: low density mat molded and low density R-RIM (Reinforced-RIM) composites, numerous thermoplastics, R-RIM fascia and body panels, etc. [6]. The regrind stock method for R-RIM compounds will be used to illustrate this technology [7–9]. This process allows the reuse of painted or unpainted R-RIM molded parts into new products such as painted bumpers and rocker panels. In this process the R-RIM scrap is ground into a fine powder and mixed before processing into one of the reactive components.

Grinding is a two step process: material from sprues, manifolds, rejected parts (painted and unpainted) or sorted, washed parts from scrapped automobiles are granulated in a knife mill to a particle size of approximately 3 mm, and then, in the second step pulverized in an impact disk mill to powder with a particle size of < 250 μm. This mill can be run under a full recycle setup to reduce all the scrap polymers to less than 250 μm, but in this case the feed rate must be reduced.

The cost of the grinding process has a great effect on the economy of the regrind method. It depends on the throughput, required particle size, the hardness of the R-RIM material, the level of utilization of the unit and several other factors. The finely ground R-RIM powder can now be used as inorganic filler at about 10% by weight without significantly changing the physical-mechanical properties of the finished part; only the elongation at break is reduced slightly (Table 1).

In the production of PUR-RIM systems extended by DETDA (diethyltoluene diamine), when the RIM regrind is added to the polyol system containing amine, it absorbs the DETDA leading to uncontrolled variations of the mixing ratio and changes in mechanical properties. The subsequent diffusion of the absorbed amine to the surface of the finished product causes staining of its painted surface. The problem can be solved by using a three component RIM process in which the

Table 1. Physical properties of a DETDA system with 15% by weight of glass fibers, with and without ground R-RIM stock [9].

	Proportion of Regrind	
Property	0% by Weight	10% by Weight
Density (kg/m³)	1180	1200
Hardness (Shore D)	59	63
E-Modulus in blending (kPa)	900	836
Elongation at break (%)	160	133
Tensile strength (kPa)	28	24.6

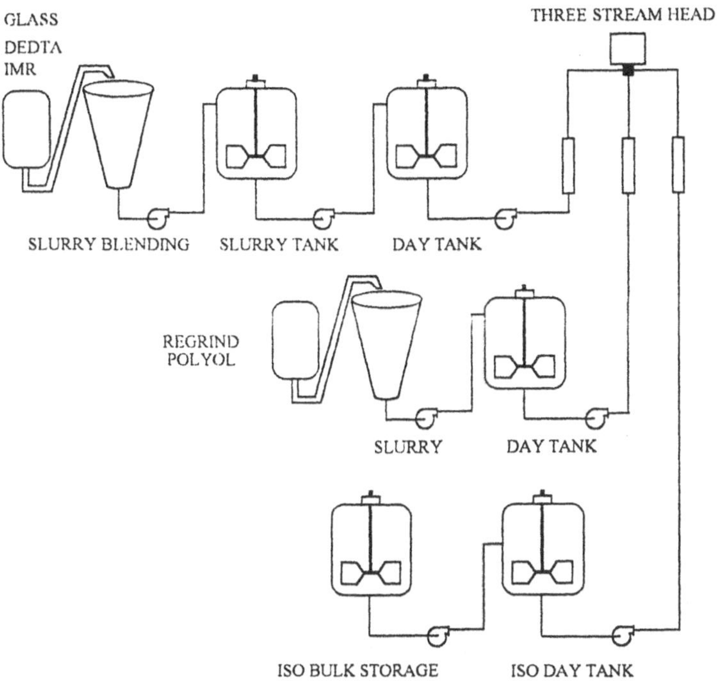

Figure 1. Flow sheet for the three component process [9].

DETDA is separate from the component containing regrind (see Figure 1). The first component is a mixture of regrind with a polyether polyol. The second component contains the DETDA, the activators and the IMR (internal mold release) along with the necessary glass fibers. The isocyanate constitutes the third component.

Various combinations of scrap materials from painted and unpainted RIM and R-RIM fascia and body panels have been reduced to the regrind form and processed into test plaques and full fascia. The three stream RIM processing of the regrind from painted RIM scrap creates no loss in surface quality [9]. The regrind particles in the RIM process are encapsulated in the new polymer in the same manner as inorganic filler such as milled glass. It can be seen that they hardly deteriorate by the addition of the regrind material. A different technique of pulverization of solid polymer wastes was realized by some researchers [10]. The solid state shear extrusion (SSSE) method is based on creating a large free surface during simultaneous application of high pressure and shear deformation. With pressure, shear strain and temperature at optimum conditions, the total amount of a solid polymer can be successfully converted into powdered material of high dispersity. This principle was originally applied to thermoplastic wastes. Recently, this technology was successfully applied to pulverization of thermoset polyurethane foams.

Pulverization of PUR foams to a very fine powder is possible over a wide range of processing temperatures, e.g., 20–100°C. The finest powder is obtained for low density flexible PUR foams and has a size range of 17–32 μm. The biggest size is for rigid PUR foam which has a size range of 96–170 μm. The shape of the powder is fiber-like with an axis ratio of approximately 1:2, which is independent of rigidity (modulus of elasticity) of the PUR foam matrix. The powder is used as a reinforcing agent for different composites based on PS and PVC matrix. The main drawback related to application of this principle is a high energy consumption.

Flake and Particle Rebonding

Flake bonding is a method where small pieces of foam are rebonded with isocyanate. It is used extensively in the slabstock industry. Recovered polyurethane flexible foams already attained a good market outlet, for example as carpet underlay. Just in the United States alone more than 200,000 ton/year are used for this purpose [11]. The demand is so great that foam wastes from Europe are imported to supplement the availability of foam scrap in the U.S.A.

In Europe, this method was adopted as a recycling route for rigid PUR foams for the manufacture of building panels of high density.

The Society for the Development of Plastics Recovery in Germany is also looking into the rebonding of particles obtained from composite polyurethane products from the car industry, taking advantage of the robustness of the particle bonding process.

Hot Compression Molding

The principle of hot compression molding is based on the findings that granulated materials made of polyurethane and polyurea, when subjected to elevated temperatures (185–195°C), elevated pressures (30 to 80 MPa) and high shear force, can coalesce and build up new bonding forces [12–14].

This technology was introduced by Bayer AG in about 1988, and appears to be particularly suitable for the recycling of RIM polyurethanes. These polymers are segmented block copolymers in which the hard segments are physically crosslinked while the soft segments are chemically crosslinked. The polymer contains bonds (hydrogen bonds) which can be split reversibly under certain processing conditions. Under the combination of the high compressive forces and the high temperature, the hydrogen bonding is disrupted and polymer flow can occur. On cooling, the hydrogen bonding in the hard domains reform and hold the new polymer in its new shape.

For compression molding of granules, several factors, such as correlation between the particle size and geometry of the regrind, the shear forces during the process, as well as the nature of the PUR used and the process parameters, must be considered. However, the smaller the particle size used, the better is the resulting surface of the newly manufactured part. The process is based on producing waste or used parts which are chopped to a screen size of 2–3 mm with an appropriate cutting device. Granulation is carried out at ambient temperature and "oversized" (250–1000 μm) material from the pulverizer is being used in hot compression molding.

For the molding process, which is carried out at 185–195 °C, the granules are preheated at 140–150°C in an oven for 1–12 minutes and transferred into a steel tool with shear edges. Its pronounced flowability allows use of undercut and extended vertical areas in the tool [13]. No release agent is required for demolding.

Examples of the mechanical properties of compression-molded articles from granules containing 22.5% by weight milled glass fibers are shown in Table 2.

It should be noted that the compression molding process produces moldings with the same density as the original material, the same hardness and only slightly lower tensile strength, whereas the elongation at break of recycled material made from painted parts drops to about 30% of the original value.

Table 2. Mechanical properties of articles compression-molded from Bayflex 110/80 granules [14].

Properties: (Painted/Non-Painted)	Original Ptd./Non-Ptd.		Compression Molded Ptd./Non-Ptd.	
Density (kg/m³)	1260	1220	1260	1260
Hardness (Shore D)	68	67	66.5	65–67
Tensile strength (MPa)	25	26	19	18
Elongation at break (%)	120	130	40	75
E-Modulus (MPa)	1100	1300	700	700
Impact, −25 °C (kJ/m²)	8.5	15	9	7

This relatively simple technique will permit 100% recycling of PUR, RIM and R-RIM, particularly when the formulations of these systems have been optimized for recycling.

Chemical Recycling

The aim of chemical recycling is the recovery of defined chemical raw materials from polyurethanes, to be reused in production.

Because of the current waste disposal situation, with landfills closing and difficulty obtaining incineration permits, a production scale chemical process for recycling polyurethane scrap is justified.

The different chemical processes, such as hydrogenation, hydrolysis, alcoholysis (glycolysis), aminolysis and pyrolysis have been known for a long time and still are studied by industry, universities and other scientific organizations.

Among these chemical processes, glycolysis has been studied extensively and it is presently being used on an industrial scale.

Hydrogenation

Many bonds of polymers can be cleaved by hydrogenation, e.g., C–C bonds of polyolefins, esters, urethanes, etc. Most types of plastic can be degraded in this way and therefore hydrogenation can be used for mixed plastics without sorting them.

Since the reaction proceeds very slowly, hydrogenation is usually carried out at elevated temperature and pressure, in the presence of cobalt-molybdenum catalyst. The yields of liquid hydrocarbons are high (>60%) [15].

A basically well-known process taken up again by Union Rheinische Braunkohlen Kraftstoff AG involves the recovery of wastes with plastic content. The basic principle of the hydrogenation process is shown in Figure 2.

The hydrogenation is carried out in a reactor in which the temperature is a maximum of 500 °C and pressure reaches about 40 MPa. The reactor content is heated in the absence of oxygen with a high hydrogen surplus. Even oxygen, which may possibly be released from the waste, is immediately bound by hydrogen, with the result that the catalytic activity of the metals is conserved.

The hydrogenation stages are followed by a lamination zone and a separation of the products into solids, gases and liquids.

When a plastics blend is used, about 65% oils, 17% fuel gas and 18% solid residue are obtained. The liquid fraction can subsequently be

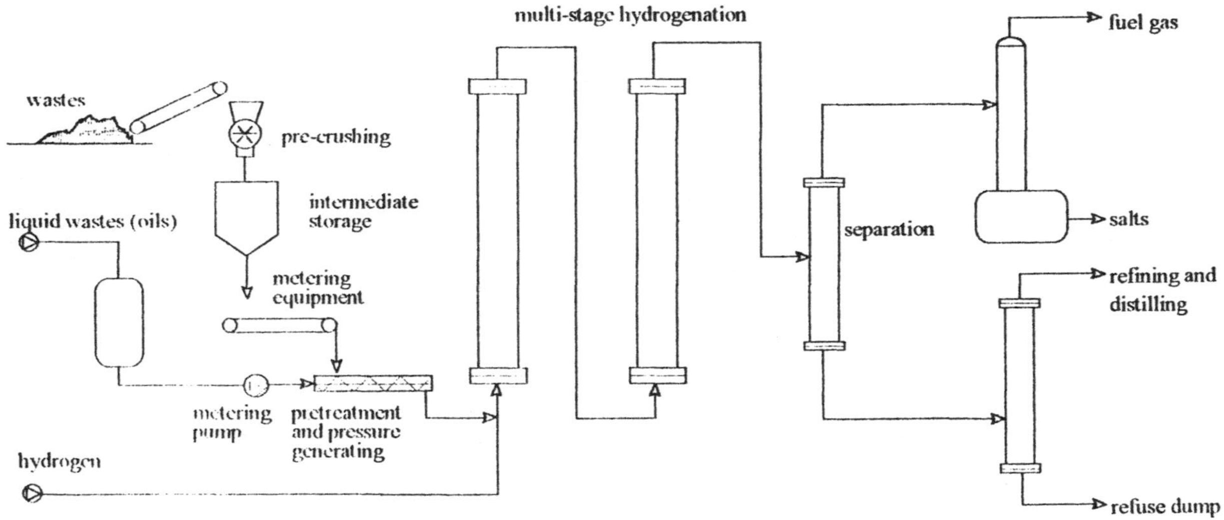

Figure 2. Flow diagram of the hydrogenation of plastics waste [15].

processed by the usual refinery methods, e.g. into gasoline, fuel oil or other products. The exhaust gases formed during hydrogenation are scrubbed and neutralized. They can be used as fuel gas. The solid residue must be disposed of.

In this process, heteroatoms such as chlorine, nitrogen and oxygen present in the long-chain molecules of polymers, are split off and the intermediate products are saturated by hydrogen. This also prevents the formation of dioxins.

It must, however, be considered that hydrogenation has up to now been confirmed in laboratory tests as a possible process concept and the general opinion is that years of work will be needed before hydrogenation can be scaled-up to accept plastics as starting materials.

Hydrolysis

Many plastics, manufactured by condensation reactions such as polyamides, polyesters, polycarbonates, polyureas and polyurethanes, when exposed to the action of water vapor, high pressure and high temperatures can be broken down into their constituent raw materials by hydrolysis.

For polyurethanes, the reaction conditions necessary for hydrolysis were determined both in discontinuous autoclave experiments [16,17] or continuously in an extruder [18]: with a temperature between 200 and 230 °C and under pressure, it is possible to obtain a selective scission of PUR chains, by reaction with water, obtaining the starting polyol, carbon dioxide and the amine corresponding to the starting isocyanate, according to the following reactions:

urethane group:

$$-\Phi\text{–NH–CO–O–R}'\text{–} + H_2O \rightleftharpoons -\Phi\text{–}NH_2 + CO_2 + -R'\text{–OH} \tag{1}$$

urea group:

$$-\Phi\text{–NH–CO–NH–}\Phi\text{–} + H_2O \rightleftharpoons 2\text{–}\Phi\text{–}NH_2 + CO_2 \tag{2}$$

where Φ is from the isocyanate and R' from the polyol.

The hydrolysis yield can be considered quantitative under proper reaction conditions, leading to a mixture of products (amines, polyol and small amounts of ureas) whose separation into single components seems possible, at least theoretically, but involves considerable processing costs, including costs necessary:

1. To distill off the amine from the crude hydrolyzed mixture and purify the remaining polyether by washing
2. To precipitate the amine from the crude hydrolyzed mixture by adding acid, and remove the precipitate by filtering
3. For liquid-liquid extraction

The polyether yields are not largely affected by the type of formulation and are of the order of 95–98%, while amine yields are about 85–92%. The polyether polyol fraction resists the hydrolysis conditions without being changed and can be recovered as originally used; recovered amines, before being used again in the foam production, must once more be converted into isocyanate. This is possible, at least in theory, in a large scale chemical operation.

If the waste contains material based on different formulations, polyether mixtures are obtained, whose reuse is a very much more complex problem than if it contains material of the same formulation, especially in view of the fact that high molecular weight polyethers cannot be separated by distillation.

Conversion by means of a continuous process has been studied on a twin-screw extruder equipped with long, co-rotating screws [18]. This was designed to withstand temperatures of up to 300 °C and the resultant pressure that would build up, and to give a residence time for the material of between 5 and 30 minutes, depending on the throughput.

The foam is broken down into flakes or powder and evenly passed into the hydrolysis zone by a feed screw. The water required for hydrolysis is pumped in against the pressure of the system, immediately next to the compression zone. Hydrolysis starts immediately and very soon converts the solid, compressed foam material into a pulp which is thoroughly wetted by the water with the help of the kneading discs on the screw. The completely hydrolyzed material finally leaves the extruder at a temperature of about 200 °C. The liquid is removed via a controlled pressure relief system; it consists mainly of polyether and the original amine from the isocyanate that was used. The CO_2 formed during hydrolysis escapes in gas form in the pressure relief system. Figure 3 shows a schematic representation of the hydrolysis equipment.

The techniques developed for flexible PUR foams can be applied to other groups of plastics, in particular rigid PUR foams and elastomers as well as polyester, polyamides and polycarbonates.

The hydrolysis process for PUR flexible foam waste has been developed to the semi-technical level, but since wastes are currently re-

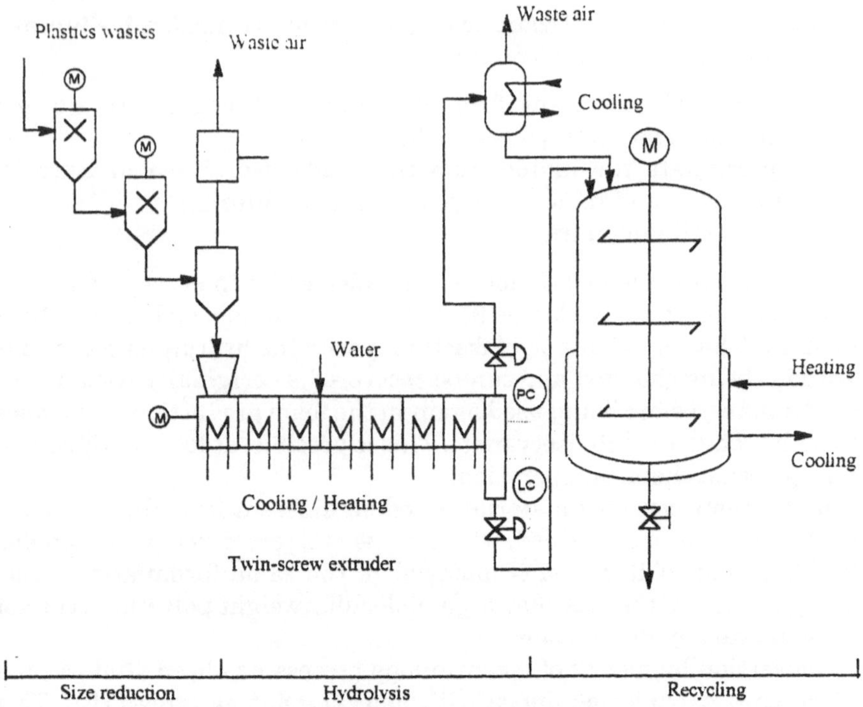

Figure 3. Continuous hydrolysis process [18].

cycled, largely in foam rebonding, hydrolysis has not so far achieved much practical importance.

Hydroglycolysis, which occurs when glycolysis is performed in the presence of base catalysts and water, gives the same products, but allows operation under easier conditions with respect to hydrolysis. The hydroglycolysis of flexible foams [19,20] with steam results in a mixture of polyols and polyamines that involves an onerous process of separation and recovery of amines with phosgenation.

The method utilizes the large changes in solubility of polyol in hydrocarbons (such as hexadecanes) that occur with changing temperatures. The high quality polyether polyol obtained can replace as much as 50% of the virgin polyol in order to produce high-quality, low density polyurethane foam. The economics of this process are not very attractive.

Glycolysis

The chemical processes up to now examined are not suitable for the solution of the problems of small and medium size industries that need to treat their wastes and scraps, also in limited quantities, locally, without adding transport costs, with an easy process and if possible, reutilize the obtained products directly in the same production lines. From this point of view a glycolysis process, a form of alcoholysis, applicable to most polyurethane wastes and scraps, seems more promising.

Glycolysis is the major chemical route so far adopted for recycling polyurethanes on an industrial scale. It essentially consists of a simple means of turning the polyurethane into a mixture of polyols which can be reused in a mixture with virgin polyol, in the same production line or in different applications.

If blown polyurethanes have to be recycled, the blowing agent must be recovered too. The removal of CFCs from PUR insulation foams is a current problem under intensive study in the recycling of rigid foams for thermal insulation. In spite of thorough physical treatment, a residual amount of CFC is retained in the foams; removal of CFC from the pretreated material during glycolysis is a possibility that is being investigated [21]. With semi-rigid and flexible foams, the glycolysis conditions are more favorable chemically than with rigid PUR, because of their lower crosslink density.

However, the flexibility of PUR chemistry makes it possible to find other uses for such polyols by appropriate admixture with virgin polyols.

Glycolysis plants are in operation in Germany, Italy, and France as well as in the USA.

GLYCOLYSIS CHEMISTRY

The chemistry of the glycolysis process involves transesterification of the carbamate, urea and ester groups in the polyurethane polymers with the glycol solvent.

Under the right conditions, polyurethane reacts with the glycol (HO–R″–OH) forming products with a low molecular weight, with terminal groups of active hydrogen, liquid at room temperature, according to the following reactions:

urethane:

$$-\Phi-NH-CO-O-R'- + HO-R''-OH \rightleftharpoons$$
$$-\Phi-NH-CO-O-R''-OH + -R'-OH \qquad (3)$$

urea:

$$-\Phi-NH-CO-NH-\Phi- + HO-R''-OH \rightleftharpoons$$
$$-\Phi-NH-CO-O-R''-OH + -\Phi-NH_2 \quad (4)$$

ester:

$$-R-CO-O-R'- + HO-R''-OH \rightleftharpoons$$
$$-R-CO-O-R''-OH + -R'-OH \quad (5)$$

where Φ comes from the isocyanate, R from the acid, R′ from polyol and R″ from the glycol.

From these reactions, a mixture of starting polyol, low molecular urethanes with −R″−OH end groups and amine compounds, obtained from urea groups, results.

At the glycolysis temperature, thermal degradation of the urethane and disubstituted ureas cannot be excluded [22–24]:

$$-\Phi-NH-CO-O-R'- \rightleftharpoons -\Phi-NCO + -R'-OH \quad (6)$$

$$-\Phi-NH-CO-O-CH_2-CH_2-R'- \rightarrow$$
$$-\Phi-NH_2 + CO_2 + CH_2{=}CH-R' \quad (7)$$

$$-\Phi-NH-CO-O-R'- \rightarrow -\Phi-NH-R' + CO_2 \quad (8)$$

$$-\Phi-NH-CO-NH-\Phi- \rightleftharpoons -\Phi-NCO + -\Phi-NH_2 \quad (9)$$

Moreover, in the presence of water, hydrolysis of the urethane and urea groups should also be considered [Reactions (1) and (2)]. Other reactions between urethane and pyrolysis products, such as amines or isocyanate, have been suggested [25].

Reaction (3), having a practically zero ΔH, is strongly influenced by the activity of the hydroxyl groups (dilution entropy) and therefore, with a high glycol/polyol ratio, at equilibrium, all the high molecular weight polyol used in the production of the foam is liberated.

The glycolysis of polyurea/polyurethanes is more complex because both urea and urethane linkages are present and the conversion of urea carbonyl to urethane carbonyl [Reaction (4)] is not thermoneutral.

From a previous study, ΔH for this reaction was estimated to be between 2 and 7 Kcal/mol [19].

Clearly, the amines coming from Reaction (4) cannot be avoided. They should, theoretically, lead to an equilibrium of aminolysis by reaction with the urethane group:

$$-NH-COO- + -NH_2 \rightleftharpoons -NH-CO-NH- + -OH \quad (10)$$

with formation of new ureas and polyureas with low solubility in the system.

The results of glycolysis depend on the following parameters:

1. Type of polymer
2. Type of glycol
3. Polymer/glycol ratio
4. Temperature of reaction
5. Type of catalyst

Type of Polymer

Glycolysis implies a transesterification reaction between urethane and urea groups of the polymer with the glycol, leading to the conversion of an even crosslinked structure to linear fragments of low molecular weight. In the reaction system the amines lead to an aminolysis equilibrium [Reaction (10)] with formation of new ureas.

Isocyanurate groups are not affected by the reaction and remain unchanged in the product.

In polymers derived from polyester polyol, the ester group undergoes a hydrolysis that starts at temperatures lower than that of the carbamates.

Type of Glycol

Low molecular weight glycols, such as MEG (monoethylene glycol) and PG (propylene glycol), yield a glycolysis product, under the same conditions, with molecular weight and viscosity lower than the product obtained by using higher MW glycols.

Their use, however, at glycolysis temperatures, which are higher than their boiling point at atmospheric pressure, requires reactors operating under pressure with the usual complications in driving this process. Hence, it is usually better to resort to higher MW glycols such as DEG (diethylene glycol) and DPG (dipropylene glycol).

The use of branched glycols, for example DPG, allows one to obtain homogeneous glycolysis products, without phase separation, even in the treatment of polymer from high molecular weight polyether polyols (ex. flexible foams). Moreover, the glycolysis product with DPG shows good compatibility with isocyanates with consequent ease in their reuse. Mixtures of glycols may also be used for various purposes.

Glycol/Polymer Ratio

In order to support the transesterification equilibrium, the glycolysis is carried out with an excess of glycol. This excess, which is present in the product as free glycol, has a solvent function for the reaction products (carbamates, ureas, amines), and solid substances at room temperature.

Upon increasing the glycol/polymer ratio, the viscosity of the final product becomes lower and the hydroxyl number rises.

It must be taken into account that these bifunctional glycols reduce the average functionality of the product with negative consequences in cases where recycling is aimed at production of polymers with high crosslink density.

Temperature of Reaction

The optimum range of temperature for PUR glycolysis, when catalysts are present, has proven to be 180–220°C. At temperatures lower than 180°C, catalysts don't turn out to be active; at temperatures higher than 220°C the rates of the secondary unwanted reactions become competitive towards transesterification reactions, leading to a product with a hydroxyl number remarkably higher than the theoretical value, in consequence of the formation of a considerable amine content. Temperatures of about 200°C have proven to be sufficient to lead to a rapid homogenization of the system and to complete glycolysis in relatively short times (3–5 hours).

Glycolysis Catalysts

The glycolysis chemistry is very complicated, both because of the presence in the polymer of numerous groups taking part in the reactions and because at the process temperature, secondary reactions cannot be completely excluded.

All these reactions have their own reaction rate that can be controlled by the action of catalysts. The use of selective catalysts for the transesterification reactions active at the lowest possible temperatures is therefore of paramount importance.

Basic substances are already mentioned in the first studies on PUR glycolysis. DEA (diethanolamine), used as coreagent with DEG for PUR glycolysis [24], increases the rate of the involved reactions, leading to products with a considerable amine content. These aromatic amines which are by-products in the glycolysis process, represent toxic and noxious substances.

Amines turn out to be present, in large amounts, also in the glycolysis products of rigid PUR foams, when K-acetate is employed as catalyst [26]. Alkaline hydroxides have been used by Braslaw [19,20] in the hydroglycolysis of flexible foams. In this process the isocyanate component of the polymer is completely converted into amines and polyureas.

When PUR scrap and wastes are glycolyzed, water cannot be excluded since glycols are not completely anhydrous and PUR absorbs water in equilibrium with the atmosphere. This water, if basic catalysts are present, leads to amine formation by hydrolysis.

Several experimental trials have been carried out by the authors [26] in order to evaluate both catalytic activity and selectivity of different substances in glycolysis.

Tables 3 and 4 make reference to glycolysis trials carried out with DPG at 190°C on model compounds of polyurethanes and polyureas, respectively. These trials have been carried out with 0.041 moles/100 g DPG of urethane (–NHCOO–) and urea (–NHCONH–) groups in a N_2 flux.

The glycolysis is incomplete even after 10 h if catalysts are absent. With a catalyst amount of 10^{-3} mole/100 g DPG, the rate of solubilization and polymer glycolysis is greatly increased.

Tables 3 and 4 show the moles of CO_2 developed for 100 moles of groups involved in the glycolysis; its determination has been carried

Table 3. Glycolysis of polyurethane foam with different catalysts.

Test No.	1	2	3	4	5
Catalyst	none	DEA[(a)]	KOH[(b)]	KAc[(c)]	$Ti(BuO)_4$[(d)]
Dissolution time (h)	12.5	2	0.5	1	0.5
Moles CO_2/100 —NHCOO—	0.91	25.6	16.5	4.5	0.4
Viscosity (mPa·s) 25°C	20.5	15.5	15.5	16.9	17.0

[(a)]DEA = diethanolamine—Aldrich.
[(b)]KOH = potassium hydroxide—Aldrich.
[(c)]KAc = potassium acetate—Aldrich.
[(d)]Titanium (IV) *n*-butoxide—Janssen Chimica.

Table 4. Glycolysis of disubstituted urea with different catalysts.

Test No.	6	7	8	9	10
Catalyst	none	DEA[(a)]	KOH[(b)]	KAc[(c)]	$Ti(BuO)_4$[(d)]
Dissolution time (h)	>10	1	0.75	0.75	0.75
Moles CO_2/100 —NHCONO—	0.94	27.9	22.15	17.0	0.36
Viscosity (mPa·s) 25°C	13.7	12.1	12.2	12.2	13.4

(a)DEA = diethanolamine—Aldrich.
(b)KOH = potassium hydroxide—Aldrich.
(c)KAc = potassium acetate—Aldrich.
(d)Titanium (IV) *n*-butoxide—Janssen Chimica.

out by absorbing the CO_2 in a $Ba(OH)_2$ solution and subsequent titration.

The data in the tables point out the fact that basic substances lead to a great evolution of CO_2 and consequently to a high amine group content (mostly primary and secondary) in the glycolysis product, which is in agreement with other researchers' data [19,20,25] which employ basic substances for the hydroglycolysis of flexible foams to amines and polyols.

Organometallic compounds such as $Ti[O(CH_2)_3CH_3]_4$ and other substances based on tin seem to be otherwise selective catalysts for the glycolysis.

The importance of the catalyst choice depends on the fact that it is impossible to have anhydrous reacting systems. The difficulty of obtaining anhydrous glycol is in fact well known. Industrial glycols are commercialized with a water content between 0.1 to 0.2% w/w. It must also be taken into account that polyurethane wastes can absorb humidity from the surroundings. Tests carried out on pulverized foams show an equilibrium at 20°C and 100% R.H. reached in 15 days with water absorption of 1.5%.

These trials have shown the possibility of carrying out glycolysis reactions of urea and urethane products also in the presence of water, keeping the hydrolysis reactions down to a minimum, and therefore limiting the formation of amines produced from secondary reactions.

Glycolysis of Rigid Polyurethane Foams

The first attempt to recycle PU foams by heating scrap foam with the polyol used to prepare the original foam was disclosed in 1960 [27]. The ratio of scrap foam to polyol was low because of viscosity limitations. Several more recent patents also claim heating of scrap foam in large excess of polyol or glycol but usually two layers were formed [28,29]. A similar result was obtained when scrap foam was heated in monoethanolamine (MEA) in the presence of base [30].

Efforts were designed to produce a homogeneous polyol, which would not separate upon standing into immiscible layers; this was achieved by heating scrap foam with glycols such as monoethylene glycol (MEG), diethylene glycol (DEG), propylene glycol (PG), dipropylene glycol (DPG), etc. in the presence of a catalyst [31–33].

Sometimes it is advantageous to use a mixture of glycols in order to achieve the desired viscosity or reaction profile.

Some experiences of glycolysis with different glycols are presented in the following. Wastes from industrial rigid polyurethane foam thermal insulators with different formulations (Table 5) were subjected to glycolysis.

Pilot plant runs were performed in a 50 dm^3 jacketed reactor, equipped with stirred and thermostatic oil circulation. The reactor was charged with a mixture of glycol and catalyst; then, the temperature was raised to the required value (210°C) and scrap was fed continuously through a screw. The rate of feeding was adjusted to the rate of dissolution.

In these trials, the catalyst/foam ratio was constant, but the type of glycol and the glycol/foam ratio were varied. The glycolysis conditions and the characteristics of the polyols produced are reported in Table 6.

Characteristics of the polyols obtained in runs 4 and 5 are also re-

Table 5. Formulation of foams subjected to glycolysis [33].

Foam Components (parts by weight)		
Sample:	A	B
Polyether polyol, sucrose-based[(a)]	—	100
Polyether polyol, amine-based	125	—
Trichloroethylphosphate	12	12
Glycerol	5	5
Water	0.5	0.5
Surfactant[(b)]	1	1
Amine catalyst[(c)]	2	2
Trichlorofluoromethane	12	37
NCO index	110	110
Oligomeric MDI[(d)]	160	160

[(a)]Hydroxyl number 475 mgKOH/g.
[(b)]Tegostab B 1048—Th. Goldschmidt AG.
[(c)]Polycat 8—Air Products.
[(d)]Tedimon 31—Enichem.

Table 6. Glycolysis runs on pilot plant scale [33].

Run Number	1	2	3	4	5
Parts by weight:					
DEG	100	100	100		
DPG	—			100	100
Foam A	50				
Foam B		75	57	57	57
Characteristics of the polyol produced:					
Specific gravity (kg/m³), 25°C	1150	1160	1160	1100	1100
Viscosity (mPa·s), 25°C	681	4121	1723	7430	7250
Hydroxyl number (mgKOH/g)	774	608	692	540	550
pH (isopropanol/water mixture), 25°C	5.5	5.8	5.8	5.7	5.6

ported in Table 7. As can be observed, the quality and the values of average hydroxyl functionality (calculated using the hydroxyl number and the Mn value) were remarkably constant.

The pH depends on the amount of flame retardant used in the manufacture of the foam. A decrease of the acid number can be achieved by neutralizing the polyol with propylene oxide or other commercially available epoxides.

The polyols obtained in pilot plant runs were blended with a fresh sorbitol-based polyol (hydroxyl number = 490 mgKOH/g) and used as raw materials for further foaming. The behavior of these polyols in foaming is practically equivalent to that of fresh polyols (Tables 8 and 9).

Under the same conditions, polyols from DPG glycolysis gave better foams than those obtained using polyols recovered by DEG glycolysis;

Table 7. Characteristics of the polyols from glycolysis runs nos. 4 and 5 [33].

Sample No.	4	5
Specific gravity (kg/m³), 25°C	1100	1100
Viscosity (mPa·s), 25°C	7430	7250
Hydroxyl number (mgKOH/g)	540	550
pH (isopropanol/water mixture), 25°C	5.7	5.6
Water (% by weight)	0.15	0.15
Total alkalinity (meq/g)	0.88	0.85
Unsaturation (meq/g)	0.098	0.100
Mean numeric molecular weight (Mn)	225	247
Mean hydroxyl functionality (*f*)	2.44	2.44

Table 8. Properties of foams obtained from blends of fresh polyol and polyol recovered by glycolysis with DEG [33].

Foam Components (parts by weight):							
Polyether polyol, sorbitol based	100	70	50	40	30	20	0
Glycolysis polyol from run no. 3	0	30	50	60	70	80	100
Properties:							
Density (kg/m^3)	28	29	30	30	29	30	30
Compression strength (kPa)	167	164	147	120	115	107	81
Thermal conductivity (mW/m·K)	19.8	19.8	19.8	20.9	20.1	20.0	20.0

moreover they have higher compressive strengths than those obtained from fresh polyols and the aromatic content of the recovered polyol increased the flame retardancy of the new foam. The good thermal conductivity indicates the fine cellular structure and high content of closed cells.

INDUSTRIAL PROCESS INFORMATION

An industrial glycolysis plant, operative in an Italian industry, consists of a 4 m^3 reactor with heat exchangers and input/output devices.

Scrap and wastes of rigid foams cut from blocks (Figure 4) are broken down and sent to collection points in silos (Figure 5). The cut-up or pulverized foam, in a weighted amount in a loading silo, is continuously fed through a hopper or a screw type solid feeder into the heated reactor containing the specified glycol at 190–210°C under a slight nitrogen purge (Figure 6).

Table 9. Properties of foams obtained from blends of fresh polyol and polyol recovered by glycolysis with DPG [33].

Foam Components (parts by weight):							
Polyether polyol, sorbitol based	100	70	50	40	30	20	0
Glycolysis polyol from run no. 4	0	30	50	60	70	80	100
Properties:							
Density (kg/m^3)	28	29	29	28	29	29	29
Compression strength (kPa)	167	171	196	206	200	186	137
Thermal conductivity (mW/m·K)	19.8	19.4	19.6	20.6	18.7	18.3	18.2

Figure 4. Scrap and wastes of rigid PUR foams cut from block.

Figure 5. Silos for collecting pulverized foams.

Figure 6. Glycolysis plant.

The rate at which the foam is fed into the reactor is a function of the type of agitation, mass and heat transfer. The foam does not wet very easily and since by nature it is very light in weight, it has a tendency to float on top of the hot glycol; as a result, good stirring is important. Foreign matter should be physically removed prior to feeding of the foam. However, small amounts of insoluble products are readily removed by filtering the polyol through an in-line filter.

The product obtained under the above mentioned conditions from glycolysis with DPG (foam/glycol ratio 58/100) was used, mixed with virgin polyol (sorbitol-based, hydroxyl number 490 mgKOH/g), using a standard formulation, for preparing new rigid foams. The properties of the foams obtained in the laboratory with the usual techniques, are given in Table 10.

The foams in Table 10 show characteristics not inferior to those of foams obtained with only virgin raw materials. By increasing the glycolysis products, the stability of the foams in humid conditions underwent a reduction as a consequence of the lower functionality of the recycled product. In industrial applications, it is better to use a recycled/virgin polyol ratio not exceeding 60/40.

The plant can treat up to 1200 kg of scrap every 8 hour working shift, showing both satisfactory ecological and economical results.

Table 10. Properties of foams obtained from blends of virgin polyol and glycolysis product.

Foam Components (parts by weight):						
Polyether polyol, sorbitol based	100	80	60	40	20	0
Glycolysis product	0	20	40	60	80	100
Properties:						
Density (kg/m^3)	28.0	27.8	27.4	28.0	27.9	27.5
Compr. strength (kPa)	162	180	172	176	170	165
Dimensional change (Δ1%)						
70°C, 95% RH, 7 days	1.76	1.46	1.80	1.85	1.97	2.05
70°C, 7 days	0.57	0.50	0.75	0.78	0.82	0.90
Thermal conductivity (mW/m·K)	19.8	19.4	19.6	20.3	20.0	19.2

Glycolysis of Flexible Polyurethane Foams

Flexible PUR foams, water blown, are mostly made of polymers with repeating urethane and urea groups and therefore they are apt to undergo a degradation process by glycolysis.

Today industrial scrap and post-consumer flexible foams from car seats are recycled by flake bonding using isocyanate adhesives, resulting in products for carpets and packaging. It is an economical process but limited mostly by the lack of market requirement.

Studies on chemical recycling via glycolysis of flexible foams are described in the literature [34–37]. Preparation of the wastes is the first stage in any waste management process aimed at recovering polyurethanes. Apart from the characteristics of the waste, the environmental and economic impact of transporting and handling low density PUR foam waste greatly affects the viability of recycling unless the foam can be densified at the source.

Cryogenic pulverization of flexible foams requires about 1.5 kg of liquid N_2/kg PUR [36] and is thus expensive; other authors [38] report that cryogenic pulverization of flexible polyester urethane foams could be performed more efficiently and economically by incorporating an intermediate densification step into the process. In this case, liquid nitrogen was reduced to 1 kg per 1 kg of output, by careful optimization of operating parameters.

ICI Polyurethanes [37] has developed a compaction process based on a standard design of a pellet mill in cooperation with Amandus Kahl Nachf. The technology of the pellet mill incorporates two steps in the case of polyurethanes:

- pulverization of foam to sub-cell size particles
- compaction of the resulting powder into granules

The first step is achieved by feeding foam flakes between rapidly moving steel rollers and the die plate. The second step is achieved by pressing the powder into die holes where they can agglomerate according to the process conditions and the pellet mill design.

Trials of reduction in volume of flexible foams from car seats have also been set up by EniChem using two different techniques [34]:

- densification of foams starting from whole seats
- micronization through extrusion at low temperature of foam crumbs

Experimental work produced powders having the characteristics summarized in Table 11. The table shows that it is possible to obtain powders having particle sizes ranging from 100 to 1000 μm with different distribution.

Particularly interesting is the densification procedures based on a special milling technique which shows the following advantages:

- use of entire seats without previous grinding
- obtainment of materials of different sizes as a function of processing time
- a larger productivity and lower energy consumption in comparison with conventional grinding systems

An extremely flexible continuous/discontinuous process for volume

Table 11. Techniques for reduction in volume of flexible foams [34].

Densification

Type	Time (min)	Size % (μm) >1000	710	300	125	Average Size (μm)
1	30	22	39	33	6	600
2	60	10	10	65	15	385

Extrusion

Type	Output Rate (kg/h)	Temp. (°C)	Size % (μm) 710	300	125	Average Size (μm)
1	3.4	60	15	50	35	214
2	9.2	60	45	45	10	260
3	16.5	60	35	45	20	242

and size reduction of flexible foams has been achieved in order to fulfill recycling needs; its general scheme is shown in Figure 7.

These powders and pellets have been used as feedstocks in the glycolysis process obtaining a series of glycolyzed products (polyols) which can be employed as raw materials in the automotive field. Characteristics of the products are summarized in Table 12. Of considerable concern are products obtained with the use of a co-glycolizing medium Q (not specified) which allows us to:

- obtain products having low viscosity and relatively low hydroxyl content
- generate derivatives having low alkalinity and consequently a reduced free aromatic amine content

The choice of a glycolizing medium can influence the reapplicability as follows:

- MEG is suggested as glycolizing agent for products to be used in the preparation of materials having high flexural modulus related physico-mechanical properties.
- DPG and a co-glycolizing agent are suggested for products to be reused in materials having elastomeric characteristics.

Moreover, in order to obtain homogeneous polyols it is essential to use glycols containing branched methyl groups (example DPG).

The use of MEG allows the process to be carried out with high polymer/glycol ratio (up to 4/1) but a by-product is obtained with a top liquid phase, mainly formed by the polyether polyol from the polymer and a bottom phase formed by the solution of carbamates, ureas and aromatic amines in MEG. The TDA amine content (2,4–2,6 toluene diamine) is higher than that acceptable for safe handling of the product without using special precautions.

The possibility of transforming free aromatic amines by means of reaction with formaldehyde into products with methylol end groups has been studied [39]. As a formaldehyde (CH_2O) source, HMTA (hexamethylene tetramine) was chosen because at the glycolysis temperature it reacts, forming CH_2O and NH_3. It is believed that the reaction products of the HMTA react with TDA isomers that are formed in the system, not allowing the formation of the aminolysis equilibrium products. A material results which has a free amine content lower than the detection limit of the method (about 100 ppm).

ICI Polyurethanes [37] has developed a split phase glycolysis process as a recycling technology for MDI-based flexible foams (Figure 8).

This is a technology particularly suited to flexible foam slabstock

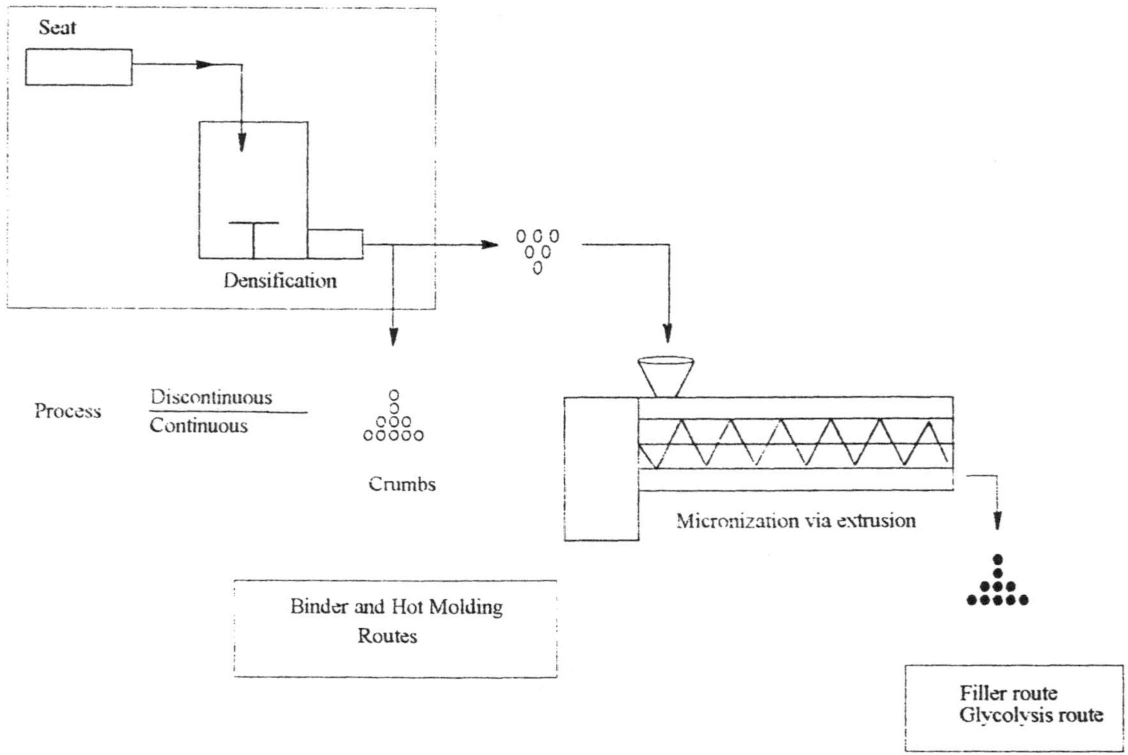

Figure 7. New technology layout: densification and micronization [34].

Table 12. Chemical recycling—types of glycolysis polyols [34].

		Polyol Properties			
PU Type	PU/Glycol Ratio	OH Value, mgKOH/g	Alkalinity, mgKOH/g	Viscosity, 25°C, mPa·s	Application
Seat	75/25 MEG	573	93	3000/8000	S-RIM, Rigid Foams, R-RIM
Seat	40/60 DPG + Q	350	33	1400	Septum/underbody
Seat	50/50 DPG + Q	296	50	3900	Septum/underbody
S-RIM	75/25 MEG	780	107	1940	S-RIM
1′ Gen. RIM	70/30 MEG	690	139	1480	3′ Gen. RIM

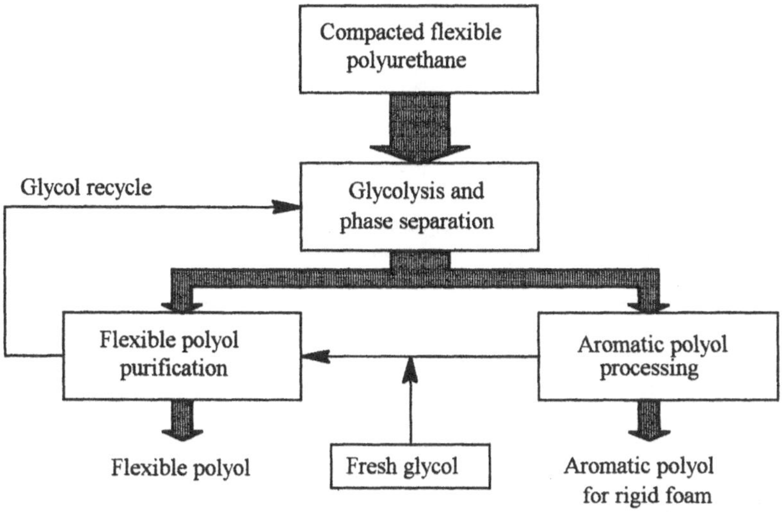

Figure 8. Split phase glycolysis simplified process scheme [37].

markets where the volumes of production wastes are sufficient to allow economies of scale in the chemical process and, as previously described, suitable compaction technology for the foam waste exists. The developments in split phase glycolysis have been targeted to obtain flexible polyols of high quality and consistency to replace virgin polyols.

The flexible polyol obtained by a split phase glycolysis process can be used for complete replacement of the original polyol normally used for prepolymers in the "Waterlily" production (Table 13).

Foams made with prepolymers using recycled polyol show improved compression set properties over those made with the original polyol (Table 14).

In accordance with the authors, the improvement is due to the removal, during purification, of low functionality and low molecular weight species present in the original polyol. This occurs during glycolysis and subsequent purification of the flexible polyol.

Split phase glycolysis yields an equal quantity of flexible and aromatic polyols. Evaluation of the aromatic polyol has shown that it can be used at up to 9% in a rigid foam lamination system without significant loss of physical properties.

Glycolysis of RIM Polyurethanes

RIM (Reaction Injection Molding) polyurethane/polyurea polymers

Table 13. Split phase recycled flexible polyol specification for prepolymers [37].

	Unit	Recycled Polyol	Specification
Appearance		Clear	Clear
OH no.	mgKOH/g	27	26–32
Viscosity	mPa·s, 25°C	1332	1050–1400
Water content	%	0.03	max. 0.08
Acid value	mgKOH/g	0.01	max 0.1
DADPM*	ppm	300	max. 500
Glycol	%	0.2	max. 0.25

*Diaminodiphenylmethane.

Table 14. Comparison of foam properties using recycled and original flexible polyol [37].

	Unit	Method	Recycled Polyol	Original Polyol	Recycled Polyol	Original Polyol
Density	kg/m^3		32.0	31.0	45.0	45.0
Mean tear	N/m	ISO/DIS 8067	153.0	152.0	164.0	180.0
Comp. set:						
Dry 50%	%	ISO 1856A	3.9	7.6	2.3	3.3
Humid 50%	%	ISO 1856C	10.9	21.2	2.7	4.0
Hysteresis	%	ISO 3386/1	34.0	36.0	27.0	30.0
CLD* 40%	kPa	ISO 3386/1	4.2	4.0	4.9	5.4
Comfort index (CLD 65%/CLD 25%)			3.8	3.9	3.5	3.5

*Compression load deflection.

are materials that, because of their intrinsic properties, find increasing use in the automotive industry (bumpers, fascias, side panels, door interior panel, etc.).

Among the different chemical processes, glycolysis of DETDA (diethyltoluenediamine) extended RIM polyurethanes has been studied by many authors [34,40–42] with the purpose of obtaining a mixture of oligomers sufficiently similar to a polyol to be used in the production of new RIM.

In the DETDA extended RIM there is a prevalence of urea groups in the polymeric chain and in this case it has been observed [40] that the urea groups, in the presence of an excess of glycol (HO−R″−OH) undergo glycolysis according to the following reactions:

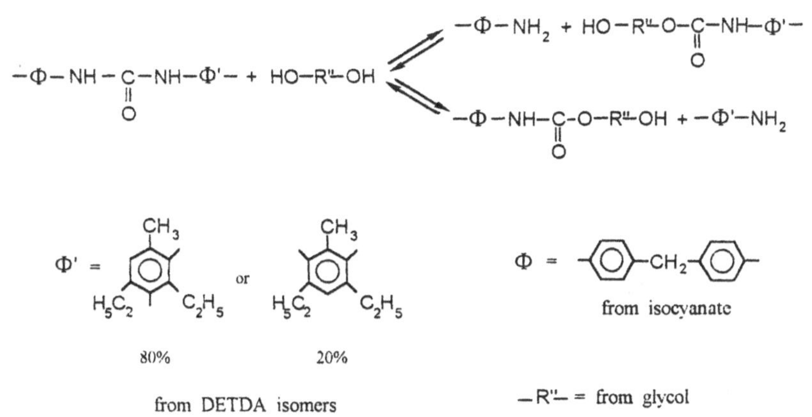

Due to the symmetry of the diarylureas present in the polymer, the glycolysis forms both DETDA and DADPM (diaminodiphenylmethane) amines.

A GPC chromatogram of the glycolysis reaction products (200 °C, 3 hours and DPG/R-RIM ratio = 1) is given in Figure 9. It shows the presence of 11 components whose identification is reported in Table 15.

The R-RIM glycolysis products present a certain number of components, principally carbamates and ureas both with low molecular weight, soluble in an excess of DPG and in the polyether polyol. The milled glass fibers of R-RIM tend to sediment with the cooling of the glycolysis products. Hot filtering (190 °C) separates the fibers. The residual amount of glass in the filtrate is about 0.1% by weight.

Some properties of the glycolysis products are given in Table 16.

The properties given in Table 16 indicate that the products could possibly be used in the preparation of new highly crosslinked polymers

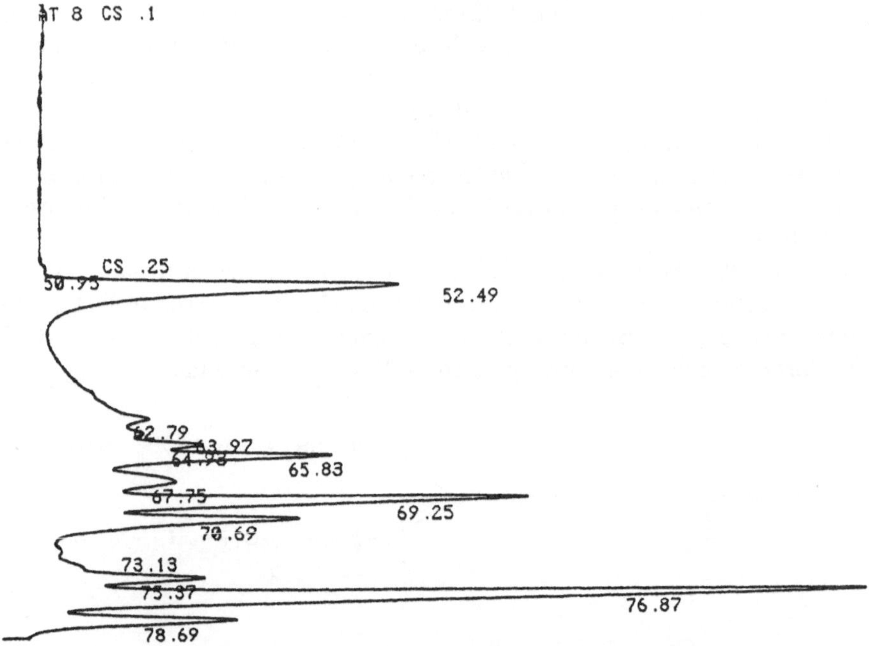

Figure 9. GPC of glycolysis products from PUR-RIM polymer [40].

similar to those used in the preparation of rigid thermal insulation foam. Rigid foams with valid properties have been produced during laboratory tests, carried out with this purpose. The residual content of glass fibers does not allow one to use these products in normal industrial plants for rigid foam production. The content of milled fibers however is not a problem if the glycolysis product is used in plants producing R-RIM.

This mixture has a high hydroxyl number and therefore cannot be used as such for the production of polymers with a suitable structure for RIM. The possibility to use it for this purpose, in substitution of the original triol, depends on the possibility of obtaining a product with a lower hydroxyl number.

The problem has been solved by some researchers [40,41], partially substituting the free DPG in the glycolysis products with a high molecular weight triol.

The basic idea of this approach is to use excessive amounts of glycols and then distill off the excess glycol. In order to offset the viscosity, virgin polyether polyol is added to the reaction mixture. The chopped RIM or R-RIM (Reinforced RIM) is transferred into the reaction vessel which

Table 15. Peak assignment for RIM glycolysis products [40].

Peak No.	RT (min.)	Component
1	78.69	DETDA isomers
2	76.87	DPG
3	75.37	DADPM
4	70.69	$H_2N—\Phi—U—\Phi'—NH_2$
5	69.25	$C—\Phi—NH_2$
6	67.75	$H_2N—\Phi'—U—\Phi—U—\Phi'—NH_2$
7	65.83	$C—\Phi—C$
8	64.98	unidentified
9	63.97	unidentified
10	62.97	unidentified
11	52.49	polyol (triol)

Where U is the urea group and C is the carbamate group of DPG.

contains the preheated glycol and, preferably mono- or dipropylene glycol. The parameters used for the glycolysis are reported in Table 17 and the flow sheet of the process is shown in Figure 10.

Similar to other glycolysis technologies it is not necessary for the new process to separate painted and unpainted PUR parts. Moreover filled or unfilled parts or products from PUR formulations with internal mold release do not create a problem.

This new process offers remarkable advantages, e.g., an optically clear stable, one phase solution with low DADPM (diaminodiphenylmethane) content and low viscosity. A GPC chromatogram of a product obtained in this way is shown in Figure 11.

As can be seen from the chromatogram, the components of the product are the same as that found in the glycolysis of the RIM with DPG. The high molecular weight triol ($\approx$6000) does not interfere with the products and simply behaves as a solvent for a homogeneous system.

Table 16. Properties of R-RIM glycolysis products (after filtration of milled fibers) [40].

R-RIM/DPG Ratio (w/w)	100/171	100/100
OH no. (mgKOH/g)	570	490
Viscosity (mPa·s, 25°C)	1970	3600
Total alkalinity (mgKOH/g)	31	44
DETDA (% w/w)[(a)]	0.8	1.5
DADPM (% w/w)[(a)]	0.8	1.6
DPG free (% w/w)	56	not specified

[(a)]Gas chromatographic analysis.

Table 17. Reaction parameters for glycolysis of R-RIM polyurethanes [41].

Size of RIM particulate matter (mm)	≅3
Starting weight ratio PUR/glycol	1:1
Glycol preheating temperature (°C)	200
Reaction time (h)	2
Reaction temperature (°C)	200
Weight ratio PUR/added virgin polyol	1:1
Post reaction time during distillation of excess glycol (h)	4–6 (140 °C/50 Torr)
Total reaction time (h)	6–8

The polyol mixture from the new glycolysis process can be reused in the original R-RIM application. Up to 60% by weight of these recycled polyols (including the polyether polyol for dilution) can be added to fresh polyol without affecting processing or properties (Table 18). The data quoted in the table shows a slight loss in properties when glycolysis products are employed.

A different series of glycolyzed products (polyols) which can be

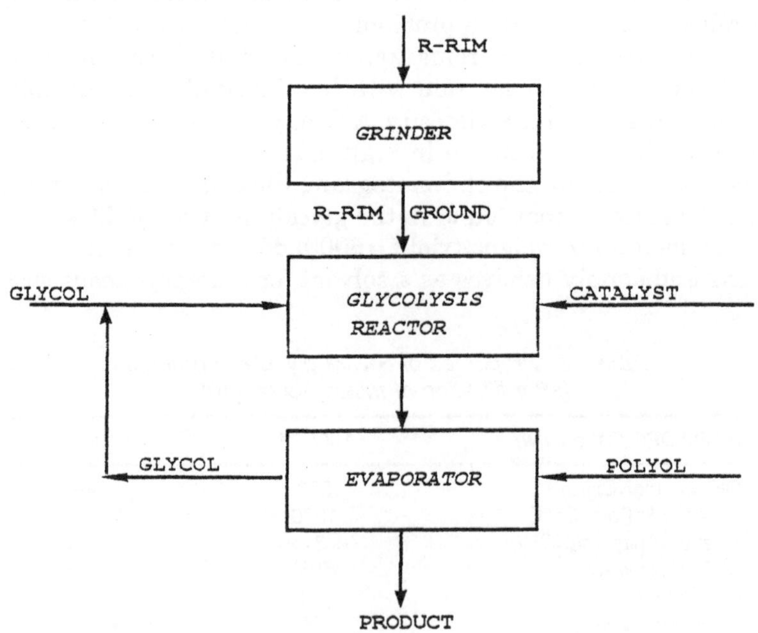

Figure 10. Glycolysis process flow sheet.

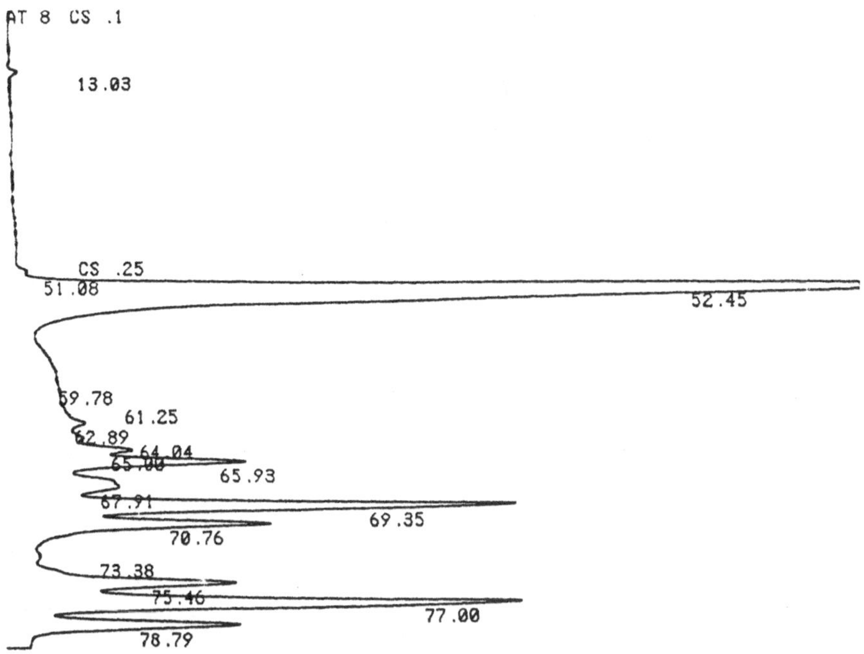

Figure 11. GPC of final product for PUR-RIM recycling [40].

employed as raw materials in the automotive field has been produced by EniChem [34], using the previously described procedures. Characteristics of the products are summarized in Table 12. Concerning structural RIM (S-RIM) polyurethanes made from recycled polyols, either from ground seats or from S-RIM materials, an optimization of formulation leads to engine undershields having physico-mechanical characteristics shown in Table 19, in line with those produced with virgin polyols. These results were achieved by using a formulated polyol having 50% of recycled polyol and an isocyanate/polyol ratio of 1.9.

Recycling of Microcellular Polyurethane Elastomer Waste

Microcellular polyurethane elastomers (MPE) are used extensively in shoe soling. The production of these elastomers is based on the reaction between polyester polyol and prepolymers of 4,4′ diphenylmethane diisocyanate (MDI) in the presence of chain extenders. Although the material and the techniques of preparation have been improved, nevertheless the process produces scrap and waste that comprise between 3 and 4% of the used raw material.

Table 18. Glycolysis process parameters and properties of RIM-PUR, obtained with glycolysis polyols (20% milled fibers) [41].

	A: Standard[a]	B	C
Virgin polyol (pbw)	100	70	40
Glycolysis polyol (pbw)	—	30	60
Polyol viscosity[b] (mPa·s)	400	550	800
Injection time (s)	0.8	0.8	0.8
Polyol temperature (°C)	55	60	60
Isocyanate temperature (°C)	40	40	40
Mold temperature (°C)	60	60	60
Injection pressure (bar/bar)	200/190	200/190	200/190
Green strength	good	good	good
Density (kg/m³)	1.28	1.29	1.27
Shore D hardness	68	67	69
Elongation (%)	164	175	168
Tensile strength (MPa)	28.1	31.0	27.4
Flexural modulus (MPa)	1450	1636	1810
HDT value (°C)	152	140	132
Dynstat impact, −25°C (kJ/m²)	16	13.6	15.2

[a]Standard: Bayflex 110/80 (Bayer AG).
[b]Measured under conditions of RIM processing.

Table 19. Application of recycled polyol in S-RIM applications [34].

	Standard	Unit	100% Virgin Polyol	50% Recycled Polyol
Density	ISO 1183	kg/m³	1.415	1.316
Tensile modulus	ASTM D 638	MPa	4256	3302
Tensile strength	ASTM D 638	MPa	57.8	39.9
Tensile elongation	ASTM D 638	%	2.1	2.2
Flexural modulus	ISO 178	MPa	2255	3227
Flexural strength	ISO 178	MPa	32	70
Impact resistance	ASTM D 256	J/m	304	347
Hardness Shore D	ISO 868		84	71
HDT (264 psi)	ISO 75	°C	49	73
Glass content	ISO 3451	%	16	14
Total fillers content	—	%	35	38

For the first time the possibility of recycling MPE scrap and waste with a glycolysis process for the production of rigid polyurethane foams [43–45] has been examined.

Both DPG and DEG have been used in the glycolysis, but with DPG, the reaction, under the same operating conditions, is slower and the product is more viscous. For the preparation of foams, products obtained from elastomer/DEG ratio of 60/40 have been therefore used. These products that have, as a result of the glycolysis conditions (200 °C, 3 hour, and catalyst), an hydroxyl number ranging from 440 and 540 mgKOH/g and a viscosity of about 1.0–2.0 Pa·s are compatible with fresh polyol of high functionality and result in a stable blend. The foams produced with the glycolysis product improved the compression properties even at low percentage of recovered polyol as a consequence of an improved cellular structure. The effect on cellular structure is also shown by the thermal conductivity; for the foams obtained with recovered polyol it decreases more than 10% as a consequence of a higher number of closed cells and their smaller diameter.

With increasing amounts of glycolysis product, the foams show lower dimensional stability. This is a consequence of the low functionality of the product and of the consequent reduction of the crosslink density. However, the functionality of the products can be improved by carrying out the glycolysis in the presence of trifunctional hydroxyl compounds [44]. This fact allows a larger use of the glycolysis products in the recycling. The use of oligomeric isocyanate with high functionality allows similar improvements.

Products from the glycolysis of MPE have also been used as polyol components in the production of polyisocyanurate foam with good properties [45].

The processes for the production of rigid foams allow the recycling of the elastomeric wastes in industries that work in different fields. Moreover, the large amounts of elastomeric scrap cannot be totally absorbed by the rigid foam industries. Therefore the opportunity of recycling elastomeric wastes in the production of new elastomers [46] was examined.

The problems that can be found in this case are linked with the particular physical structure of the PUR elastomers which must not be altered by recycled material.

The microcellular elastomers present a dual phase structure in which domains of rigid segments act as crosslink points for the flexible matrix. This biphasic structure formed during the polymerization reaction is due to the incompatibility of aromatic isocyanate with aliphatic polyols. The polyols used for the elastomers have molecular weights of about 2000 and are obtained by condensing an aliphatic diacid with glycol: MEG, DEG or higher MW glycols with 2 to 3% of TMP (trimethylolpropane).

In the glycolysis process, it is impossible to obtain homogeneous products with similar values of molecular weight. It was therefore thought better to carry out the glycolysis with a high glycol/elastomer ratio and thereafter increase the molecular weight by a condensation reaction with a diacid.

Glycolysis has been carried out on a sample whose formulation is given in Table 20. The optimal conditions for reducing secondary reactions are a temperature of 190–200 °C, the presence of $Ti[O(CH_2)_3CH_3]_4$ ($1.5 \cdot 10^{-3}$ mole cat./100 g elastomer) as catalyst, reaction time of 3 hours. The elastomer/glycol ratio was varied between 40/60 and 60/40. Some of the characteristics of the glycolysis products from the elastomer of Table 20 under the conditions mentioned above, are given in Table 21.

The properties of the products given in Table 21 suggest the possibility to use them as chain extenders in the preparation of new elastomers.

Unfortunately in this way only a partial recycling of the wastes is possible and it does not resolve the problem of disposal. Therefore the possibility was examined to prepare, with the products obtained, compounds with high molecular weight by condensation of the glycolysis products with a diacid in order to obtain products that can be used as polyol components for elastomer production.

Table 20. Elastomer PUR model [46].

Components (parts by weight):	
Polyester polyol[(a)]	100.00
MEG (monoethylene glycol)	13.60
Water	0.45
Catalyst[(b)]	0.45
Surfactant[(c)]	1.00
	115.50
Prepolymer[(d)]	115.50
NCO index	100.00

[(a)]Polyester polyol obtained from adipic acid (55 w/w), DEG (42 w/w) and TPM (3 w/w); slightly branched, hydroxyl number 57 mgKOH/g.
[(b)]DABCO 33 LV—Air Products.
[(c)]Rhodorsil 3199—Rhone Poulenc.
[(d)]Prepolymer with 21.5% NCO, obtained from pure 4,4′ diphenylmethane diisocyanate (MDI) (68.3 w/w) and polyol[(a)] (31.7 w/w).

Table 21. Characteristics of the MPE glycolysis products [46].

Product	1	2	3
Elastomer/DEG ratio	40/60	50/50	60/40
Viscosity (mPa·s)20 °C	310	1220	1940
Density (kg/m^3)20 °C	1.170	1.183	1.195
Hydroxyl number (mg KOH/g)	641	550	470
—NH_2 content (meq./g) × 10^3	2.3	3.2	4.3
Mn (from GPC)	225	314	348
Mw (from GPC)	320	490	604
Mw/Mn	1.42	1.56	1.74
MEG (%)	3.7	2.6	1.8
DEG (%)	26.1	17.1	9.2

The glycolysis products have been esterified with adipic acid (AA) by condensation in an inert atmosphere (N_2), starting at a temperature of 160 °C, increasing it up to 200 °C and operating for the final part under a vacuum (20 mmHg). Total time for the condensation was about 8 hours. The choice was not to use condensation catalysts because they require neutralization before use. The amine content of the products is lower than that of the glycolysis products, i.e more than what is due to dilution with AA, suggesting the possibility of amidation (Table 22).

The products shown in Table 22, mixed in different ratios with fresh reactants, have been used for the production of new elastomers. The resulting products have been compared with the equivalent ones obtained from virgin products.

The results for the three series of tests are given in the following tables. In the first series, the esterification product has been mixed with 50% fresh polyol (Table 23), following the formulation given in Table 20, but without the blowing agent (H_2O). The second series of tests (Table 24) has been carried out using as the polyol, only the es-

Table 22. Esterification of glycolysis products and properties of esters [46].

Test	A	B	C
Glycolysis product	1	2	3
Elast./DEG/AA ratio (% w)	24.1/36.2/39.7	32.3/32.3/35.3	42.2/22.1/29.7
Hydroxyl number (mgKOH/g)	59	64	68
Acid number (mgKOH/g)	0.4	0.2	0.9
—NH_2 content (meq./g) × 10^4	3.4	4.2	5.1
T_g (°C)	−53	−48	−46

Table 23. Elastomers from mixtures of esterification products and virgin polyol (50/50 w/w) [46].

Properties	Ref.	A	B	C
Density (kg/m³)	1.102	1.129	1.132	1.095
Tensile strength (N/mm²)[a]	10.89	12.84	13.72	12.38
Elongation at break (%)[a]	353	333	330	341
Tear resistance (Nmm)[b]	73.8	79.7	85.0	79.0
Hardness (Shore A)[c]	86	88	88	87
Abrasion resist. (mg/mm³)[d]	129	130	141	140

[a]ASTM D 638.
[b]ASTM D 642.
[c]ASTM D 2240.
[d]DIN 53516.

terification product and the prepolymer of Table 20. The third series of tests (Table 25) has been carried out using exclusively the esterification product.

The analysis of the results given in Tables 23–25 allows one to consider the following points: by examing the series of data from the same ratio of elastomer/glycol, it can be observed that by increasing the esterification product, the elastomeric properties decrease.

This is particularly evident for the elongation at break. This decay increases from products A to C. This fact seems to be attributed to the quantity and distribution of aromatic rings in the polyester chain. In product A these seem more diluted in the chain, separated by flexible segments longer than in the other products. This situation seems to favor the formation, during the phase of the elastomer preparation, of a biphase structure, which is responsible for the elastomeric properties of these materials [47].

Table 24. Elastomers from esterification products substituting virgin polyol [46].

Properties	Ref.	A	B	C
Density (kg/m³)	1.102	1.145	1.165	1.032
Tensile strength (N/mm²)[a]	10.89	11.91	13.56	8.53
Elongation at break (%)[a]	353	292	334	266
Tear resistance (Nmm)[b]	73.8	81.1	83.1	70.7
Hardness (Shore A)[c]	86	88	88	87
Abrasion resist. (mg/mm³)[d]	129	118	146	153

[a]ASTM D 638.
[b]ASTM D 642.
[c]ASTM D 2240.
[d]DIN 53516.

Table 25. Elastomers from only esterification products [46].

Properties	Ref.	A	B	C
Density (kg/m³)	1.102	1.135	1.087	1.047
Tensile strength (N/mm²)[(a)]	10.89	11.40	10.38	11.92
Elongation at break (%)[(a)]	353	254	220	213
Tear resistance (Nmm)[(b)]	73.8	75.3	67.5	71.6
Hardness (Shore A)[(c)]	86	88	84	88
Abrasion resist. (mg/mm³)[(d)]	129	113	170	161

[(a)]ASTM D 638.
[(b)]ASTM D 642.
[(c)]ASTM D 2240.
[(d)]DIN 53516.

By increasing the amount of aromatics, the phase separation decreases, forming materials which are more rigid and less resistant to abrasion.

Figure 12 shows the relationship of the percentage of elongation to the total content of aromatics in the polymer.

From the profile it can be deduced that an increase in the aromatics amount of 4–5%, with respect to that of standard elastomers from fresh

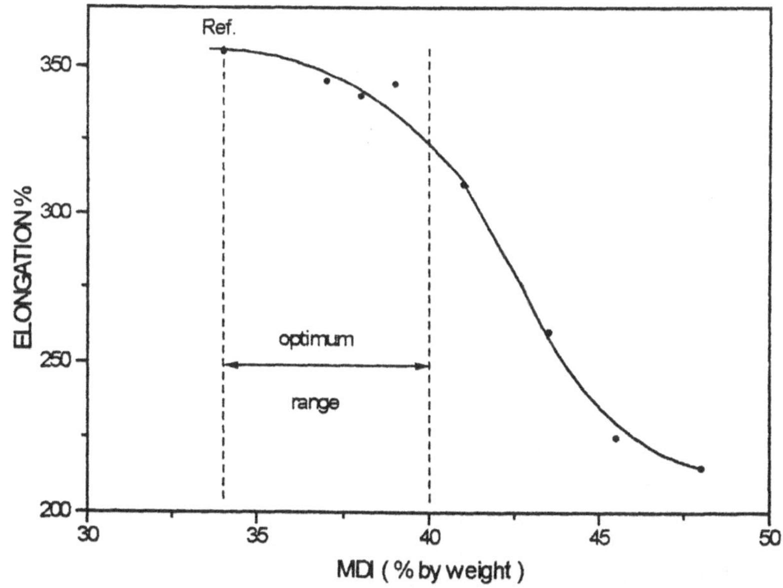

Figure 12. Total content of aromatics in polymer vs. elongation % [46].

material, is acceptable, allowing total recycling of the wastes and scraps.

Pyrolysis

Pyrolysis is a tertiary recycling technology that involves a thermal cleavage process, applicable to all plastics produced from hydrocarbons that can be converted into petrochemicals. Pyrolysis is concerned with the thermal decomposition of organic materials at higher temperatures in which oxygen may completely be excluded.

It deals with low-temperature cracking that yields gases, condensable products and solid residues. In some cases, e.g., polystyrene and poly(methylmethacrylate), it is possible to recover the monomers [48].

The advantage of pyrolysis over combustion is a 5 to 20 fold reduction in the volume of product gases which leads to considerable saving in gas purification [49].

Depending on the products obtained, pyrolysis may be considered material or energy recycling. One of the greatest problems is connected with the fact that polymeric wastes are poor thermal conductors and consequently, the degradation of macromolecules requires considerable amounts of energy.

A lot of processes and reactors have been studied (Table 26); the best results have been obtained using rotary kilns and fluidized beds.

Pyrolysis is an endothermic reaction in which heat must be added. The heat needed for pyrolysis can be supplied directly (oxygen-air) or indirectly (via heat-exchange media). In the latter we have two separated chambers, a combustion chamber and a pyrolysis chamber. Increasing the pyrolysis temperature reduces the solid residue, the content of water, and the light oil fractions. At the same time there is an increase in the gas yields, which have high caloric value [50].

The particle size and uniformity of the feed to a pyrolytic reaction has a considerable effect on the reaction rate and the distribution of the resulting products. The finer the particle size of the feedstock, the faster are the reaction rates, the larger the gaseous yields and the smaller the yields of solids and liquids. When pre-drying takes place, the pyrolysis reaction needs less heat supply for the reactor, requiring less heating time, and produces a smaller liquid fraction [51].

The wastes have relatively long residence times in rotary kilns (≥ 20 minutes), whereas residence times in fluidized-bed reactors rarely exceed a few seconds, with a maximum of 1.5 minutes [49].

Due to the large temperature gradient in the rotary kiln, the product spectrum is very diverse. For this reason gases and oils resulting from the process are normally used for the direct generation of energy (heat).

Table 26. Processes for pyrolysis of plastics wastes [49].

Process/Developer	Procedure	Products	Development Stage
Mitsubishi	melting vessels, indirectly heated, 500°C	oil, some gas	100 kg/h Tokyo, Japan
Union Carbide	extruder barrel, indirectly heated, 420–600°C	waxes	pilot plant, USA
Kobe Steel	rotary kiln, indirectly heated, 500–700°C	oil, gas, energy	1 t/h, Kobe, Japan
Voest-Alpine*	shft kiln, directly heated	gas, steam	650 kg/h, pilot plant, Linz, Austria
KWU process*	rotary kiln, indirectly, 450–500°C	energy	3 t/h, plant, Ulm, Germany
Noell, Dr. Otto process	rotary kiln, indirectly heated, 650–700°C	pyrolysis oils, gas	6 t/h, plant, Salzgitter, Germany
Waterloo University	short-dwell pyrolysis, 500–700°C	oil, gas	laboratory plant, Waterloo, Canada
West Virginia University*	two fluidized beds, one of which is oxidizing	gas	laboratory plant, West Virginia University, USA
Tsukishima, Ebara*	two fluidized beds, one of which is oxidizing	energy	4 t/h, Yokohama, Japan
Hamburg process	fluidized bed, indirectly heated, 600–900°C	pyrolysis oils, gas, carbon black	20–60 kg/h, University of Hamburg; 0.5 t/h, ABB plant, Ebenhausen Ingolstadt, Germany

*Plastics mixed with household garbage.

Fluidized bed pyrolysis has turned out to be particularly advantageous because of the high heat and mass transfer as well as constant temperature throughout the reactor, so that the product spectra is largely uniform and moving parts in the hot zone of the reactor are absent.

The reactor works at a temperature of 600 to 900 °C, while the pyrolysis gas, used to create the swirl in the fluidized bed, is preheated to 400 °C by indirect exchange through radiative pipes that are heated by pyrolysis gas.

Table 27. Feeds and products from a fluidized-bed pyrolysis of plastic waste [48].

Feed	Pyrolysis Temp. (°C)	Gas (weight %)	Oil (weight %)	Residue (weight %)	Others (weight %)
Polyethylene PE	760	55.8	42.4	1.8 C	
Polypropylene PP	740	49.6	48.8	1.6 C	
Polystyrene PS	580	9.9	24.6	0.6	64.9 Styrene
PE/PP/PS	750	52.0	46.6	1.4	2.1 H_2O
Polyester	768	50.8	40.0	7.1	5.0 H_2O 0.3 HCN
Polyurethane	760	37.9	56.3	0.5	3.4 HCN
Polyamide PA-G	760	39.2	56.8	0.6	2.5 H_2O
Polycarbonate	710	26.5	46.4	24.6	97.2 MMA
Poly(methyl methacrylate)	450	1.25	1.4	0.15 C	56.3 HCl
Poly(vinyl chloride)[(a)]	740	6.8	28.1	8.8	1.5 Steel
Poly(tetrafluoro-ethylene)	760	89.3	10.4	0.3	4.6 H_2O
Medical syringes	720	56.3	36.4	5.8	14.0 Metals 1.8 H_2O
Plastic from household waste separation	787	43.6	26.4	25.4	1.0 H_2O
Plastic from car shredding	733	29.9	26.7	27.6	14.0 Metals 1.8 H_2O
EPDM-Rubber[(b)]	700	32.3	19.2	47.5	1.0 H_2O
SB-Rubber[(c)]	740	25.1	31.9	42.8	0.2 H_2S
Scrap tires	700	22.4	27.1	39.0	11.5 Steel
Lignin	500	3.4	29.9	49.3	17.4 H_2O
Cellulose (Wood)	700	47.1	23.0	18.6 C	11.3 H_2O
Sewage Sludge	600	34.3	27.7	33.2	4.8 H_2O

[(a)]PVC: Poly(vinyl chloride).
[(b)]EPDM: Ethylene-propylene-diene-monomer.
[(c)]SB: Styrene-butadiene.

Table 28. Components from the pyrolysis of different plastic wastes in a fluidized-bed process [48].

Pyrolysis Products	Polyethylene, wt % (740°C)	Plastics from Household Separation, wt % (787°C)	Polyester, wt % (768°C)	Polyurethanes, wt % (760°C)
Hydrogen	0.75	0.70	0.30	0.23
Methane	23.6	17.5	3.8	3.5
Ethane	6.7	2.9	0.24	0.76
Ethylene	19.8	9.8	1.5	5.4
Propane	0.08	0.08	0.01	0.01
Propene	5.5	1.3	0.08	1.7
Butene	0.54	0.13	0.01	
Butadiene	1.6	0.44	0.04	1.2
CO_2		2.3	17.2	14.4
CO		1.2	27.4	8.7
HCN				0.31
HCl($CaCl_2$)		7.2		
Hexene	0.02	0.002	0.02	0.2
Benzene	19.1	13.8	18.3	5.0
Toluene	3.9	5.6	2.5	1.8
Styrene	0.5	3.1	2.1	0.01
Indene	0.25	1.0	0.41	1.1
Naphthalene	2.8	3.8	1.4	1.5
Methylnaphthalene	0.63	0.84	0.25	0.69
Diphenyl	0.30	0.41	1.47	0.15
Fluorene	0.20	0.22	0.18	0.27
Phenanthrene	0.52	0.66	0.08	0.62
Phenylnaphthalene	0.08	0.36	0.21	0.01
Fluoranthene	0.08	0.26	0.02	0.01
Acetonitrile				0.58
Acrylontrile				0.95
Benzonitrile				1.8
Ketones, acetals			0.15	9.2
Acetophenone			1.4	0.3
Other compounds	11.3	9.0	11.7	34.1
Carbon black	1.75	6.4	7.1	0.5
Water		2.6	2.1	5.0

The gases obtained, after passing through a heat exchanger, are separated from residual carbon and fine dust in a cyclone and then cooled to room temperature by quenching in a condenser. Subsequently they are largely stripped of liquid products, freed from small droplets in an electrostatic precipitator, compressed by membrane compressors and then stored.

Almost 40% of these gases are recycled to heat the reactor (energy recycling), some part is used to fluidize the bed, and the remainder can be used as fuel gas with a caloric value 20% higher than that of natural gas [15].

As can be seen in Tables 27 and 28, when polyurethane is used, the best pyrolysis temperature is 760 °C and the basic products formed are 37.9% (w/w) gases, 56.3% oils and the remainders are solid residues, which have to be disposed of. The oil obtained can be recycled (material recycling) to produce polymers rather than for heating.

ENERGY RECOVERY

Combustion

In common with most plastics, polyurethanes have an appreciable energy content, 28–32 MJ/kg, some portion of which is recoverable in an appropriate combustion plant. Energy recovery is only one option for polyurethane recycling, but it must be taken into account, in particular for end life materials or commingled plastics where other technologies represent only partial solutions to waste management.

Energy recovery eliminates the cost and the ecological problems of landfilling scrap, allowing us to make use of their energy content, which is high enough to allow plastic wastes to be used in garbage-based power stations to produce electricity, steam, or remote heating [52].

In modern incinerators, where combustion is quite complete and the concentration of chemicals is very low, the environmental impact is one-third to one-sixth that of a landfill site processing the same quantity of wastes [53].

Recovering energy from plastic materials is however not free from problems concerning:

- selection of the materials to feed the kilns
- choice of the kiln to employ and its operating temperature
- purification of the combustion products

The feed must be subjected to a size standardization operation in order to have the correct feed with elimination of inorganic compounds. Non-combustible matter is removed to reduce solid residues and fly ash and, at the same time, to increase the caloric value of the waste being incinerated.

Fluidized bed combustors with fluidized silica sand and rotary kiln combustors with afterburners, working at a temperature up to 1200°C, have been employed in several trials, assuring 99.99% destruction of organic constituents.

Energy recovery is carried out in a multipass, radiation type boiler which lowers the gas temperature to about 260°C, generating steam. Particulate and acid gases are removed in an appropriate multi-stage scrubber. In Figures 13 and 14 are shown the plant schemes of a fluid bed combustor and a rotary kiln combustor, respectively.

Many energy recovery or incineration trials of various polyurethanes were previously carried out by individual corporations or industry associations to determine feasibility and environmental acceptability. Trials on rigid polyurethane foams have been made by Bayer AG, employing a rotary kiln and by Thyssen Engineering GmbH in a fluidized bed combustor thus demonstrating that these materials can be completely combusted, including the CFC-11 contained as blowing agent. According to the authors, by-products of combustion, such as CO, NO_x and HCl were present below permissible levels. Due to the extremely low bulk density of the rigid foam, some modifications had to be made during the trials in the fluidized bed combustor to avoid carry over of incompletely combusted particulate matter. Automotive shredder fluff which contains the organic, polymeric fraction but also a very significant non-ferrous inorganic fraction, was also converted to energy in trial runs at Bayer AG and by Dow Chemical in conjunction with Voest-Alpine.

The results of different combustion trials carried out on R-RIM scrap coming from automotive bumpers and fascias were reported by Baumann et al. [54] in order to compare RIM scrap as a fuel source to other common industrial fuels.

In addition to the fact that society sees incineration as the "destruction of a material" and is therefore opposed to it, this form of recycling cannot be supported from the point of view of costs. The requirements for incinerating plastics are very strict with respect to limiting values for emissions, which means that the investments for the plants require three times as much capital as those for plants for material recycling [55]. This involves extra cost which in addition to the costs for collection, transport and for dumping the inert residue, makes the process

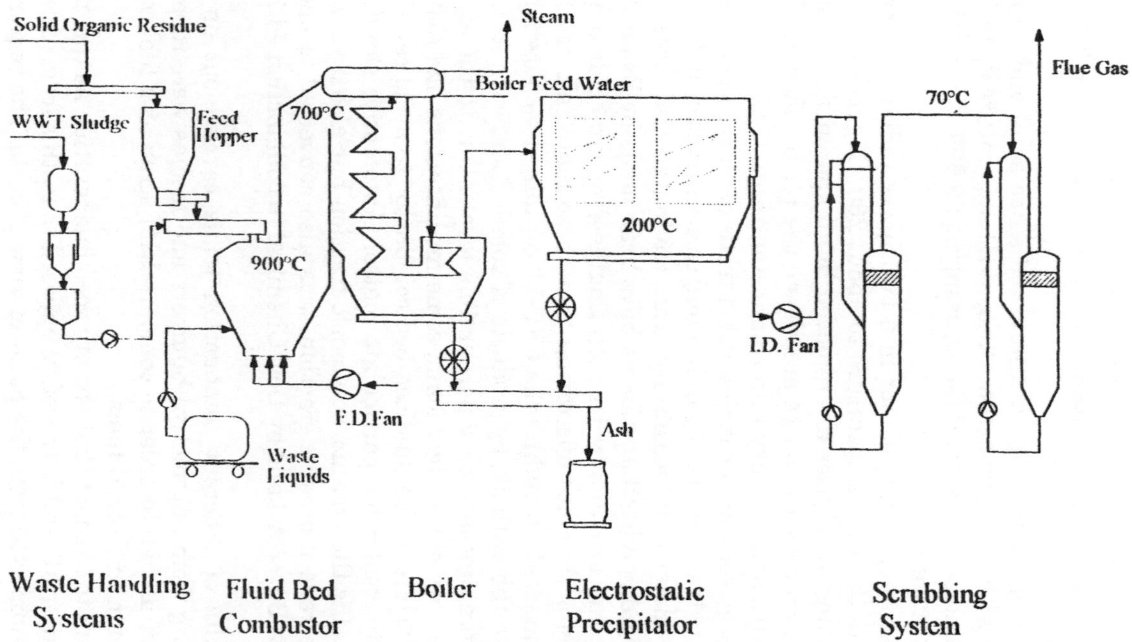

Figure 13. Flow sheet of a fluid bed combustor [54].

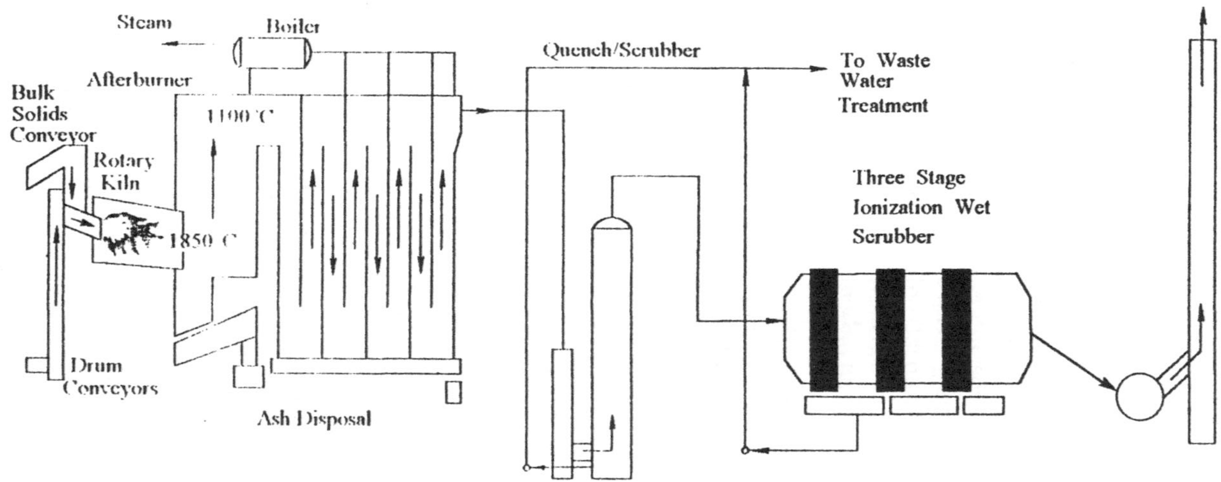

Figure 14. *Flow sheet of a rotary kiln combustor [54].*

impractical. To become economical, an incinerator needs to handle around 200,000 to 250,000 tons of refuse a year [56].

CONCLUSIONS

The recycling of plastics has to contribute to the conservation of resources and to the reduction of environmental pollution. The idea that all thermoplastics are recyclable, while crosslinked materials are not recyclable has been largely overcome.

Many options for thermoset polyurethane recycling have been presented, including physical and chemical methods as well as energy recovery. Some solutions have been developed into pilot plants, while a few are already available in actual practice.

A life cycle analysis will define the preferred criteria for selection among the different waste management technologies; the choice of material or energy recovery methods also takes into account the economical point of view.

This kind of option guarantees the applicability of recovered polyurethane in many industrial branches and offers the best possibility for the future of polyurethanes.

To recycle implies a new philosophy of production, i.e., to take greater care in choosing the materials to be employed, which is referred to as designing for recycling, and to conform the production process to the requirements for reusing scrap and end life products.

REFERENCES

1. 1994. *Plastics Handbook,* Ed. by *Modern Plastics,* p. 82.
2. 1991. "Standard Guide for the Development of Standards Relating to the Proper Use of Recycled Plastics," *ASTM D 5033.*
3. *Ullmann's Encyclopedia of Industrial Chemistry, 4th Ed.,* A 21:57.
4. Oertel, G. and W. Reichmann. 1992. *Kunststoffe,* 82(4):22.
5. Schaper, H. A. 1994. *Polymer Recycling,* 1(1):23.
6. Farrissey, W. J., R. E. Morgan, R. L. Tabor and M. Zawisza. 1992. In *Emerging Technologies in Plastics Recycling,* G. D. Andrews and P. M. Subramanian, eds., *A.C.S. Symposium Ser. 513,* p. 272.
7. Morgan, R. E. and J. D. Weaver. 1991. *International Congress and Exposition,* Society of Automotive Engineering, Detroit, MI, February 25–March 1, paper 910580.
8. Morgan, R. E., G. H. Dean, R. L. Tabor and M. Zawisza. 1991. *Polyurethanes World Congress,* September 24–26, Nice, France, p. 653.

9. Brückner, H., U. Frank, H. Fransen, W. Raßhofer, H. Schaper and H. U. Schmidt. 1991. *Kunststoffe*, 81(9):3.
10. Arastoopour, H., G. Ivanov and F. Shutov. 1993. *Cellular Polymers, 2nd International Conference*, Edinburgh, March 23–25, paper 18.
11. Hooper, J. F. G., G. Parrinello, A. Parfondry and K. W. Kroesen. 1992. *Utech '92 Conference*, The Hague, the Netherlands, March 31–April 2.
12. Meister, B. and H. Schaper. 1990. *Kunststoffe*, 80(11):13.
13. Raßhofer, W., U. Liman and J. Wagner. 1991. *Polyurethanes World Congress*, Nice, France, September 24–26, p. 640.
14. Müller, P. and R. Riess. 1992. *Die Makromolekular Chemie, Macromolecular Symposia*, 57:175.
15. Menges, G., B. Eysmondt, A. Feldhaus and H. Offergeld. 1988. *Kunststoffe*, 78(7):573.
16. Wyandotte Chem. Corp., U.S. patent 3441616, 1966 (L. C. Pezzini and J. T. Patton): *Chem. Abstr.* 71 (1969) 4107d.
17. Ford Motor Co., U.S. patent 4316992, 1982 (J. L. Gerlock, J. Braslaw and J. Albright): *Chem. Abstr.* 96 (1982) 218707g.
18. Grigat, E. 1978. *Kunststoffe*, 68(5):281.
19. Gerlock, J., J. Braslaw and M. Zimbo. 1984. *Ind. Eng. Chem. Process Dev.*, 23(3):545.
20. Braslaw, J. and J. Gerlock. 1984. *Ind. Eng. Chem. Process Dev.*, 23(3):552.
21. Ascough, M. R. and K. J. Cicalo. 1991. *Polyurethanes World Congress*, Nice, France, September 24–26, p. 361.
22. Grassie, N. 1982. *Polym. Eng. Sci.*, 22:1057.
23. Grassie, N. and M. Zulfigar. 1978. *J. Polym. Sci., Polym. Chem. Ed.*, 16:1563.
24. Ulrich, H., A. Odinak, B. Tuker and A. A. R. Sayigh. 1978. *Polym. Eng. Sci.*, 18:844.
25. Gerlock, J. L., J. Braslaw, L. R. Mahoney and F. C. Ferris. 1980. *Polym. Sci., Polym. Chem. Ed.*, 18:541.
26. Simioni, F., M. Modesti and S. A. Rienzi. 1991. *Cellular Polymers, International Conference*, London, UK, March 20–22.
27. Ten Broeck, T. R. and D. W. Peabody. U.S. patent 2,937,151, 1960.
28. McElroy, W. R. U.S. patent 3,300,417, 1967.
29. Kinoshita, O. U.S. patent 3,632,530, 1972.
30. Motsudaira, N., S. Muto, Y. Kubota, T. Yosnimoto and S. Saeto. U.S. patent 3,404,103, 1968.
31. Frulla, F. F., A. Odinak and A. A. R. Sayigh. U.S. patent 3,738,946, 1973.
32. Simioni, F. Italian patent 1,218,710, 1982.
33. Simioni, F., S. Bisello and M. Tavan. 1983. *Cellular Polymers*, 2:281.
34. Petrone, A., S. Greco and S. Chinellato. 1993. *Polyurethanes World Congress*, Vancouver, Canada, October 10–13, p. 113.
35. Simioni, F. and M. Modesti. 1993. *Cellular Polymers*, 12:337.
36. Kroesen, K. W. and D. A. Hicks. 1993. *Cellular Polymers, 2nd International Conference*, Edinburgh, UK, March 23–25, paper 16.

37. Hicks, D. A., M. Krommenhoek, D. J. Soderberg and J. F. G. Hopper. 1994. *Cellular Polymer,* 13:259.
38. Sims, G. L. A. and M. W. Angus. 1994. *Plast. Rubb. Comp. Process Appln.,* 22(2):99.
39. Modesti, M., F. Simioni and R. Munari. 1994. *6th International Conference on Polymer Supported, Reactions in Organic Chemistry (POC '94),* Venice, Italy, June 19–23, p. 224.
40. Simioni, F., M. Modesti and S. A. Rienzi. 1993. *Cellular Polymers–2nd International Conference,* Edinburgh, UK, March 23–25, paper 14.
41. Raßhofer, W., H. A. Freitag, D. I. Klier, U. Liman, K. Munzmay and J. Wagner. 1992. *UTECH '92,* The Hague, the Netherlands, March 31–April 1/2.
42. Bauer, G., D. Auchter and C. Brückner. 1993. *Kunststoffe,* 83(4):11.
43. Simioni, F., M. Modesti and S. A. Rienzi. 1987. *Cellular Polymers,* 6:27.
44. Simioni, F., M. Modesti and G. Navazio. 1988. *Macplas International,* p. 155.
45. Simioni, F., M. Modesti and C. A. Brambilla. 1989. *Cellular Polymers,* 8:387.
46. Modesti, M., F. Simioni and S. A. Rienzi. 1992. *Journal of Elastomers and Plastics,* 24(4):288.
47. Camberlin, Y. and J. P. Pascault. 1984. *J. Polym. Sci., Polym. Physics Ed.,* 22:1835.
48. Kaminsky, W. 1992. In *Emerging Technologies in Plastics Recycling,* G. D. Andrews and P. M. Subramanian, eds., *A.C.S. Symposium Ser. 513,* p. 60.
49. Kaminsky, W. 1992. *Ullmann's Encyclopedia of Industrial Chemistry,* A 21:64.
50. Steffensen, O., O. Franck, R. Rahnenfures and W. Kaminsky. 1989. *Proceedings of the Inter. Conference on Pyrolysis and Gasification.* Amsterdam: Elsevier.
51. Mustafa, N. and J. Hansmann. 1993. In *Plastics Waste Management.* N. Mustafa, ed., p. 249.
52. Steiner, K. W. 1987. *Kunststoffe,* 77:954.
53. Maraghi, R. 1993. In *Plastics Waste Management,* N. Mustafa, ed., p. 369.
54. Baumann, G. F., J. I. Myers and W. J. Farrissey. 1991. *Polyurethanes World Congress,* Nice, France, September 24–26, p. 355.
55. Hauss, A. F. 1994. *International Polymer Science and Technology,* 21(6): T/15.
56. 1990. "Recycling Polymers from Scraps Products," RAPRA Confidential Technoeconomics Report, p. 57.